Cross-Industry Innovation Processes

W0037586

Tobias Hahn

Cross-Industry Innovation Processes

Strategic Implications for Telecommunication Companies

Preface by Prof. Dr. Rüdiger Zarnekow

 Springer Gabler

Tobias Hahn
Bonn, Germany

Zugl.: Berlin, Technische Universität, Diss., 2014 u. d. T.
The Open Service Innovation Process in Cross-Industry Collaboration Networks and its
Strategic Implications for Telecommunication Companies

ISBN 978-3-658-08826-2   ISBN 978-3-658-08827-9 (eBook)
DOI 10.1007/978-3-658-08827-9

Library of Congress Control Number: 2015931211

Springer Gabler
© Springer Fachmedien Wiesbaden 2015

Printed on acid-free paper

Springer Gabler is a brand of Springer Fachmedien Wiesbaden
Springer Fachmedien Wiesbaden is part of Springer Science+Business Media
(www.springer.com)

Preface

Telecommunication markets are experiencing fundamental changes. Political, technological, business and societal conditions are changing and breaking up traditional, historically grown structures and markets. Examples are plentiful and include the deregulation of telecommunication markets, the convergence of telecommunication, media and internet industries, the growing importance of software-defined networking and rapidly changing customer behavior. Competition between telecommunication companies increases and leads to new value chains, business models and products.

Rising to these challenges, telecommunication companies increasingly focus on cooperations in the area of research and development. Especially cooperations with partners, that possess complementary skills, play an important role. On an operational level, these cooperations are traditionally implemented in the form of technology partnerships or joint ventures. In recent years the concept of open innovation has gained importance in this area as well. A close cooperation with partners, suppliers and customers is regarded as a promising approach. Nevertheless, in practice open innovation is still at an early stage and the actual success of open innovation projects varies greatly.

If the innovative capacity of a single company or a single industry sector is not sufficient, cross-industry cooperations offer a promising option. In cross-industry cooperations companies from different industry sectors jointly carry out innovation activities. Even though in practice the potential of cross-industry cooperation has been recognized, there are still only a handful of successful examples. This is especially true within the high-tech and telecommunication industry.

It is therefore very commendable, that this book specifically focuses on describing cross-industry innovation projects between telecommunication companies and companies outside the ICT sector. The insights are extremely valuable from a practical industry as well as from a scientific perspective. They offer deep and detailed information about successful innovation processes and provide practical advice and recommendations for innovation managers.

Prof. Dr. Rüdiger Zarnekow

Technical University of Berlin, Chair for Information and Communication Management

ABBREVIATIONS

AAL	Ambient Assisted Living
B2B	Business-to-Business
B2B2C	Business-to-Business-to-Consumer
BMW	Bayerische Motoren Werke (BMW AG)
BPO	Business Process Outsourcing
BT	British Telecom Group Plc.
COPD	Chronic Obstructive Lung Disease
CTO	Chief Technology Officer
DSL	Digital Subscriber Line
DT	Deutsche Telekom AG
EU	European Union
FedEx	Federal Express (FedEx Corporation)
FTE	Full-Time Equivalent
GSM	Global System for Mobile Communications
ICT	Information- and Communication Technology
IT	Information Technology
KPI	Key Performance Indicator
LOI	Letter of Intent
LTE	Long Term Evolution
M&A	Merger and Acquisition
M2M	Machine-to-Machine
MoU	Memorandum of Understanding
MS	Microsoft Corporation

NCS   National Computer Systems Group

NDA   Non-Disclosure Agreement

NPD   New Product Development

NTT   Nippon Telegraph and Telephone Corporation

OBS   Orange Business Services

OECD   Organization for Economic Co-operation and Development

P&L   Profit and Loss Statement

QDA   Qualitative Data Analysis

R&D   Research and Development

SIM   Subscriber Identity Module

SingTel   Singapore Telecommunications Ltd.

SLA   Service-Level Agreement

SME   Small and Medium-Sized Enterprises

TEF   Telefónica, S.A.

Telco   Telecommunication (company) / Telephone Conference

TPP   Technological product and process

TRIZ   'Teoriya Resheniya Izobreatatelskikh Zadatch' (theory of inventive problem solving)

TSI   T-Systems International GmbH

UMTS   Universal Mobile Telecommunications System

VCI   Verizon Communications Inc.

VGEI   Vodafone Global Enterprise Inc.

VF   Vodafone Group Plc.

LIST OF FIGURES

TABLE OF CONTENTS

1 Introduction

The first chapter introduces the research topic and explains why the specific subject was chosen. It covers practical status quo and latest research development. In addition, it defines the present studies objective and describes its research setting.

1.1 Motivation and Practical Relevance

Telecommunication companies are considerably affected by competitive and cooperative relationships in the converging software, hardware, media and telecommunications industries, resulting in major strategic challenges (Hess et al. 2012; Wulf & Zarnekow 2011). The current landscape for traditional telco-services has never been more competitive. Investments in next generation products, services and infrastructure are immense, revenues from fixed-line services are falling and companies from the information technology industry increasingly provide similar, or even superior, communication services at lower costs over the internet. In order to gain an edge in the market for next-generation communication services, or by generating completely new revenue streams with attractive new products and services, telecommunication companies are searching for strategic growth opportunities and alternative ways to harness their infrastructure and the information they own (Sienel et al. 2009; Raivio et al. 2009).

Successful commercialization of technology often requires the collaboration among horizontal competitors that have different capabilities. Co-operative activities in form of R&D joint ventures or R&D collaboration is not a new phenomenon. Since the 1970s technology partnerships especially in high-technology industries, have risen rapidly, due to increased cost and uncertainty regarding R&D efforts (Trott & Hartmann 2009). E.g., in 1971 the Japanese computer industry intensively entered partnerships with extensive interaction and information sharing, to compete against US competitors (Teece 1989).

In recent years open innovation alliances have emerged as a new collaborative paradigm in the settings of strategic alliances (Brunswicker & Hutschek 2010; Rohrbeck et al. 2009). Even though the concept of open innovation is not entirely new, because third party involvement, co-development, collaborative innovation, joint ventures, innovation clusters and networks etc. have been around for many decades, intense vertical collaboration with suppliers and customers is getting more and more popular only in recent years (Gassmann, Enkel, et al. 2010; Nesse 2008). Moreover, the concept of open innovation differs substantially from traditional, inter-organization alliances, including their strategic scope and scale, governing mechanisms, member composition and evolutionary dynamics. They generally seek to enlarge the 'economic pie' and increase the overall market demand through value co-creation, rather than fighting with competitors over market shares of a 'fixed pie' (Han et al. 2012). The ability to simultaneously collaborate with suppliers, customers and internal cross-functional teams has a positive and significant impact, on project performance and an indirect effect on market performance (Mishra & Shah 2009). Open innovation activities allow to decrease costs and risks between 60 and 90 percent and at the same time to reduce innovation cycles (Gassmann & Enkel 2006). Organizations that do not supplement their internal resources and competence with complementary external resources and knowledge show a lower capability for realizing innovations (Gemünden et al. 1992). Consequently, organizations are starting to professionalize internal processes to manage open innovation more effectively and efficiently (Gassmann, Enkel, et al. 2010).

Following the open innovation approach is one solution for telecommunication companies to overcome the obstacles in the current market environment. However, across all industries open innovation approaches are in its infancy and currently follow more a trial and error approach than a professionally managed process. The variance between a best practice in open innovation and the average is huge with a generally success rates below 25 percent (Evanschitzky et al. 2012; Gassmann, Enkel, et al. 2010). In addition, incumbent organizations spend 80-90 percent of their technology budgets on upgrades, modifications, extensions and add-ons for its present product portfolio (Roberts 2007). These incremental innovations allow a continuous, reasonable growth, to foster their position in high-turnover markets, but telecommunication companies need more radical innovation to enter new markets and to persist present pressure in the ICT industry (Roberts 2007). The current telcos' assets in isolation will offer little value, but in combination with other technologies they may solve specific business needs, that may completely transform many industries by revolutionizing their business models (Nayar 2011).

80 percent of all innovation are recombinations of existing knowledge, technologies or products (Gassmann & Enkel 2006; Gassmann & Enkel 2004). In many cases existing solutions in one industry are not sufficient for recombinations to generate radical innovation, hence, especially highly competitive markets impel a growing number of organizations to look for opportunities outside their traditional markets, to bring distant knowledge, technologies or products together with their own (Brunswicker & Hutschek 2010; Gassmann & Zeschky 2008; Alam 2007; Yamashina et al. 2002). Exploration of established solutions from distant industries reduces inherent risks related to new product development by reducing uncertainty, and cross-industry collaboration allows to develop more radical innovation (Enkel et al. 2009). There are many successful examples of technological spillovers across industries, e.g., BMW's iDrive system that is transferred from the game industry, Nike's shock absorbers that is adapted from Formula One racing technology, or FedEx's overnight package delivery that is based on methods developed at Delta Airlines (Nolf et al. 2012; Enkel & Gassmann 2010).

Managers in the high-technology industry note that while the search for cross-industry innovation is not new, cross-industry assessment is just not happening (Nolf et al. 2012). Also, various discussions with managers from European and Australian telecommunication companies in the years 2011 and 2012 reveal, that open innovation approaches find its way in innovation management and new product development, but they reluctantly intensify open innovation approaches based on collaboration with partners from distant industries, because they have hardly references and need to learn the new approach. Horizontal collaboration, across industry boundaries, is still a new topic in innovation projects, both in literature and practice and we need to understand well how to manage major/breakthrough/radical innovation efforts (Gassmann, Enkel, et al. 2010; Roberts 2007).

1.2   Existing Research

Strategic alliances have been discussed in literature for many decades and from different perspectives (Trott & Hartmann 2009). In recent research literature open innovation has developed to a central topic and even though it is a young research field, it has emerged to a mainstream research area with inescapable challenges and opportunities for researchers, policy makers and organizations (Baldwin & von Hippel 2011; Gassmann, Enkel, et al. 2010; Brunswicker & Hutschek 2010). Discussions around the new paradigm mainly include combinations of ideas, knowledge and technology distributed among a network of innovating partners (Russo-Spena & Mele 2012).

There are many ways to categorize theoretical developments in the field of open innovation, such as schools of thought, actors or processes (Calida & Hester 2010; Enkel et al. 2009; Gassmann 2006; Gassmann & Enkel 2004). Lichtenthaler (2011) identifies technology transactions, user innovation, business models, and innovation markets as main research streams in open innovation literature. *Technology transactions* focus on internal organizational capabilities for inward and outward technology transfer, and in specific on R&D alliances, absorptive capacity and external knowledge exploitation (Koppinen et al. 2010; Su et al. 2009; Stevens & Dimitriadis 2004; Li & Whalley 2002). *User innovation* focus on collaboration with users in the external exploration of new knowledge and ideas and in specific the integration of users in the innovation process (Voss 2012; Wießmeier et al. 2012; Baldwin & von Hippel 2011; Lau et al. 2010; von Hippel 1976). *Business models* focus on exploiting knowledge in open innovation processes, e.g., related to intellectual property and corporate venturing (Hummel et al. 2010; Vanhaverbeke et al. 2008; Chesbrough 2006). *Innovation markets* focus on ways to facilitate interfirm technology transfer, e.g., by the means of intermediaries that facilitate technology exchange (Nambisan et al. 2012; Gassmann et al. 2011; Howells 2006). (Lichtenthaler 2011)

By analyzing recent open innovation literature, various research trends can be identified: Industry penetration – from pioneers to mainstream, R&D intensity – from high to low tech, Size – from large firms to SMEs, Processes – from stage gate to probe-and-learn, Structure – from standalone to alliances, Universities – from ivory towers to knowledge brokers, Processes – from amateurs to professionals, Content – from products to services and Intellectual property – from protection to a tradable good (Gassmann, Enkel, et al. 2010).

The trend from standalone to alliances including R&D partnerships is based on the fact that especially in high-technology industries, organizations face increasing pressure in developing complex solutions, and consequently, revert to alternative external sources of innovative competencies (Hagedoorn & Duysters 2002). Organizations enter vertical partnerships with established suppliers to explore predominantly incremental innovation and for more radical solutions they revert to horizontal organizations across established industry boundaries (Gassmann, Zeschky, et al. 2010). In respect of an open innovation approach, a systematic creation of innovation in a cross-industry context is a new phenomenon for theory and practice, and increased recent discussions suggest that the topic of cross-industry innovation has high theoretical and practical relevance (Enkel & Gassmann 2010).

The trend from products to services is characterized by a growing number of authors recognizing that even though in all advanced economies, more than half of the economy is coming from service, literature on open innovation rarely includes service innovations compared to physical product innovations (Chesbrough & Euchner 2011; Nesse 2008; Yang 2007). Innovation research has treated services merely as a special category of products, but previous empirical investigations of innovation may not fully capture the complexities of service innovation, because they are based on narrow conceptual frameworks (Ordanini & Parasuraman 2011; Baker & Sinkula 2007). Given a shift from production oriented to service oriented economies, especially regarding the composition of telecommunication companies portfolio offerings, there is a need for new academic and professional insights that contribute to success in new service development projects (Ottenbacher & Harrington 2010).

1.3 Objective and Boundaries

Even though telecommunication companies face a volatile environment that increases pressure to look outside the company's boundaries for technology and services, to develop new business models, and to introduce new technologies, practices and organizational methods into their internal processes, uncertainty can persuade organizations to hesitate implementing significant changes (Nesse 2008; OECD & Eurostat 2005). Hence, serving practical demands and current theoretical research (cf. chapter 1.1 and 1.2), the objective of the present study is to respond to telecommunication companies' challenge to conduct service innovation projects with partners from non-ICT industries, by the means of providing a practical process model. Eventually, the process model and complementary findings may serve telecommunication companies as a guideline for strategic cross-industry innovation alliances, and may contribute in the academic context in providing new insights and propositions for existing open innovation process models.

Considering the following two major constraints is necessary to fully define the scope of the objective. It narrows the scope of application and results in a more accurate process model with better applicability in specific use cases:

- Cross-industry innovation does not necessarily implicate collaboration. It may be a sole internal activity following a traditional closed innovation process, to develop a solution for a new industry or adapting a solution from a distant industry to the own one. The present study focusses on cross-industry collaboration, because generally the innovating company is forced to conduct the project as an open innovation project, due to a lack of expertise and to assure customer acceptance. Distinction between organizations within the ICT industry is getting blurred. Overall, companies continue to be more strategic in their partnerships to spread themselves across multiple market segments (Telstra Corporation Ltd. & KPMG International 2012). Consequently, the present study focusses on horizontal cross-industry innovation with partners from non-ICT-industries, because the requirement of a partner from a non-telco industry organization would not be a decent criterion, and innovating without a partner would not exploit the potential of cross-industry innovation to its full extent (for further requirements on the partner organization cf. chapter 3.2).
- Most studies in literature do not explicitly distinguish between goods and services even though their different nature requires different innovation approaches, e.g., innovation processes for pure physical goods may need to consider production lines and logistics, while pure services may perish while creation (Segelod & Jordan 2004) (cf. chapter 2.1). Consequently, the present study explicitly focusses on non-tangible service products that are easily replicable and scalable, with no manual customization necessity. This requirement allows to exclude customer specific developments, as typically anchored in the telcos' B2B entities, and focusses on pure innovation project that allow to sell the developed

service with minor modifications to other customers as well (for further requirements on the innovation projects cf. chapter 3.2).

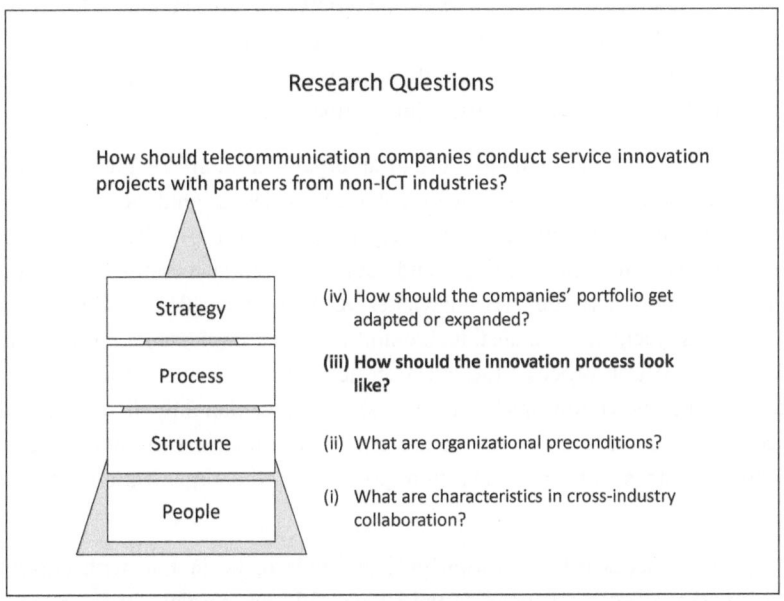

Figure 1: Research Question

To reach the main objective, (i) characteristics in cross-industry collaboration need to be conceived, (ii) organizational preconditions identified and (iv) strategic considerations deliberated, to entirely understand how to create and successfully implement the required (iii) innovation process (Roberts 2007). The three dimensions of *people*, *structure* and *strategy* are interdependent with the innovation *process* and contribute critically in achieving successful institutionalized innovation (Roberts 2007). Accordingly, the main research objective and additional considerations implicate four sub-questions that need to be answered.

Especially in radical innovation projects, where the amount of specific knowledge is high, collaboration quantity plays an important role and a partner match has a positive influence on collaboration competency (Schweitzer & Gabriel 2012; Tsou 2012a). Collaboration is fundamental for performance and innovativeness in cross-industry innovation projects (Schweitzer & Gabriel 2012; Nooteboom et al. 2007). Consequently, collaboration characteristics need to be understood, to better overcome challenges of the partners' distinct strategic, organizational, and operational contexts (Gassmann, Zeschky, et al. 2010). This cognition implies the question: (i) "What are characteristics in cross-industry collaboration?", to allow a more successful implementation of the innovation process, and consequently to contribute to the understand of how telcos should conduct cross-industry innovation projects.

Secondly, the organizational structure of an organization affects the efficiency of its innovation activities (OECD & Eurostat 2005). The alignment of internal and external structures and processes, is a key factors to improve the success rate in cross-industry innovation projects and facilitates the successful integration of open innovation activities (Brunswicker & Hutschek 2010; Chien & Chen 2010). Consequently, for a sustainable and effective implementation of a cross-industry innovation process, the question needs to be answered: (ii) "What are organizational preconditions?" Insights around the planning of organizational processes, -structures and -systems are especially important to understand what the embeddedness of the innovation process in the organizational structure implies.

Moreover, a successful innovation project needs to be in line with current or future market needs and includes the successful commercialization of the new service by its integration as a new portfolio element. Portfolio management allows senior management to implement the pursued corporate strategy (Unger et al. 2012), however, the fact that completely new services and disruptive innovation may herald a structural change, that may cannibalize the current revenue streams, poses big challenges in transferring the new service in the companies' portfolio, because it negatively influences the established personal incentive and compensation system (Chesbrough 2010, pp.94–96; Bond & Houston 2003). Hence, market needs and difficulties due to organizational rigidity and barriers need to be considered, and the question needs to be posed: (iv) "How should the companies' portfolio get adapted or expanded?"

Finally, since generations the innovation process itself is subject of many research areas (cf. chapter 1.2 and 2.3) and in focus of the present study. Its management involves the effective integration of people/staffing, planning of organizational processes, -structures and -systems and plans/strategy (Yang 2007; Roberts 2007). Consequently, in interdependence with (i), (ii) and (iv), question (iii) "How should the innovation process look like?", clarifies required activities in chronological order, to pay in the objective and main research question, of "How should telecommunication companies conduct service innovation projects with partners from non-ICT industries?" (cf. figure 1)

1.4 Research Approach

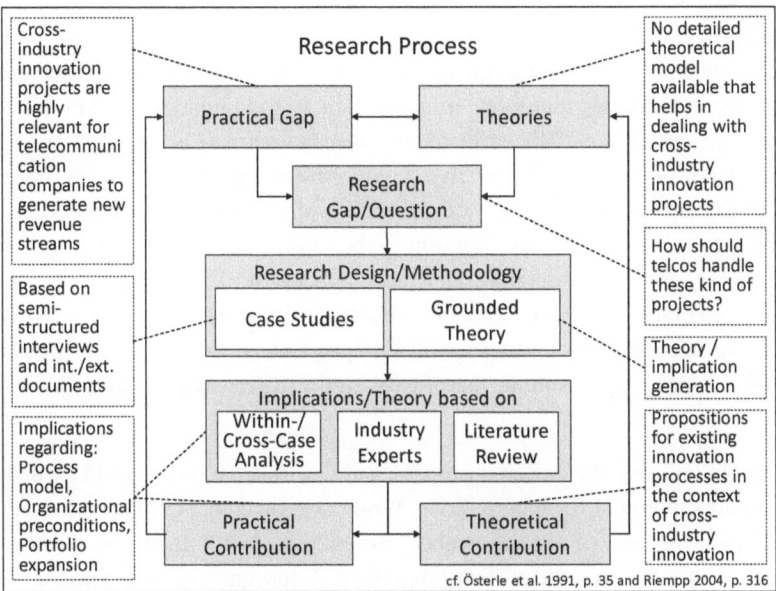

Figure 2: Research Process of the Present Study

Research is increasingly being carried out in cross-sector collaboration, linking theory and practice (Garrett-Jones et al. 2005). Since the present study's objective is to develop predominantly practical implications for telecommunication companies, the research approach is guided by the research processes defined by Österle et al. (1991) and Eisenhardt (1989).

Influenced by action research, Österle et al. (1991) define a five-stage research process that links theory and practice, to solve practical problems and generate new insights. The research process is based on a research gap, mutually defined by theory and practice, as a basis for the consequential research question.

Research subsequently structures the issue, develops and refines practical solutions, based on practical and theoretical experience, and provides practical and theoretical contributions (Österle et al. 1991, p.35). While applying the contributions in practice, research may assess and reflect them, and potentially start a new iteration while refining the research gap (Riempp 2004, pp.314–316) (cf. figure 2). Due to its relevance for practical research the suggested process is highly applicable to the present study.

To describe a roadmap for building theories from case study research, Eisenhardt (1989) combines various existing concepts, such as previous work on qualitative methods, the design of case study research and grounded theory building, and extends that work in areas such as triangulation, within case and cross-case analysis and the role of existing literature (Eisenhardt 1989). Her developed process is highly iterative, involving constant back and forth between the steps, and consists of the steps: getting started, selecting cases, crafting instruments and protocols, entering the field, analyzing data, shaping hypotheses, enfolding literature and reaching closure. Eisenhardt's approach is widely accepted and embedded in present business research and due to its extensive aggregation of concepts of data processing, the suggested process is likewise, highly applicable to the present study (Ravenswood 2011; Eisenhardt 1989).

It gets obvious that the research processes of Österle et al. (1991) and Eisenhardt (1989) are congruent for long periods. While Österle et al. (1991) emphasizes on practical proximity of the researcher, Eisenhardt (1989) focus on a complete roadmap process to guide theory building. Consequently, the present study applies both views as guiding frameworks for the research process, to build practical oriented theory from case study research, that solves current practical challenges in the telecommunication industry.

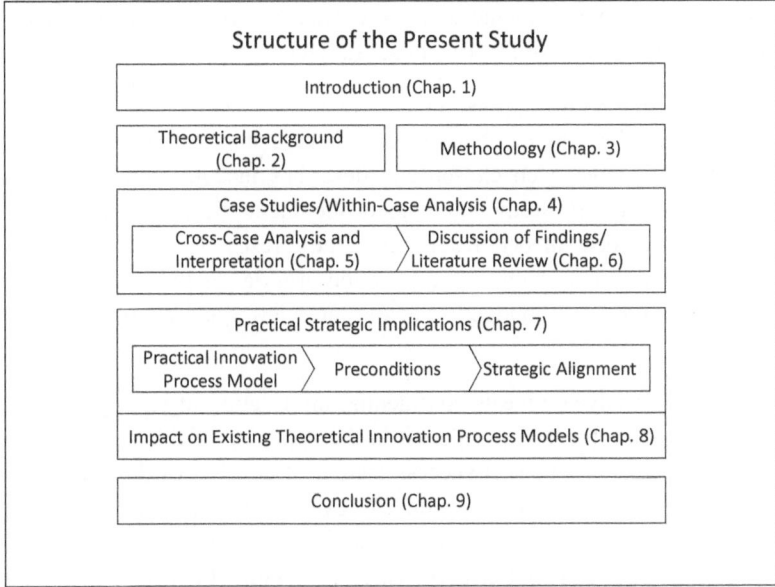

Figure 3: Structure of the Present Study

Figure 3 visualizes the structure of the present study and exemplifies how the chapters are linked. The study roughly can be divided in an introductory part covering chapters 1 to 3, theory preparation and generation in chapters 4 to 6, deduction of implications and propositions in chapters 7 and 8, and a final conclusion in chapter 9.

After a short introduction to the research topic, that emphasizes its current importance for practice and latest research development, **chapter 1** defines the present studies objective and describes its research setting.

Chapter 2 gives an overview of the theoretical background and describes the main terms of the research including service, innovation, open innovation and cross-industry innovation. The chapter differentiates between service and physical goods, and invention and innovation. To convey the evolution of innovation models, the chapter introduces the concept of innovation process generations, as one way to structure different types of approaches. Subsequently, it details structural characteristics of sequential innovation processes and emphasizes the necessity of a holistic view while analyzing innovation processes, that encompasses people, organizational processes/structures and strategy. The chapter concludes with characteristics in the ICT industry, especially related to cross-industry collaboration and innovation approaches.

Chapter 3 introduces the applied methodology and its corresponding theoretical concepts. It explains why qualitative research, case study analysis and grounded theory approach have been chosen and describes in detail and chronological order how the present study implements the methodical requirements while building a grounded theory from case study research. This part contains processual information, such as the selection criteria for the cases, creation of the interview guideline, conduction of the interviews, and analysis of the data material.

The case study analysis reveals confidential information of strategic relevance. Consequently, **chapter 4** is positioned in the appendix of this study, to protect companies' secrets and market success factors. It consists of three case studies with five innovation projects conducted by a European telecommunication company with innovation partners from Energy, Automotive and Health industry and describes their innovation approaches in detail. The within-case analysis focusses on the innovation process. As a requirement of a holistic integrated view, comprehensive information about cross-industry collaboration, competences, project and company specific organization and processes, also have been gathered.

Chapter 5 complements the previous within-case analysis by a cross-case analysis. It describes first findings, similarities and differences, across all projects. Again the analysis of the innovation process is in focus and gets enriched by the dimensional characteristics of the categories process, structure, people and strategy. The findings comprise various preliminary implications for the realization of cross-industry innovation projects and influence the emerging theory to a great extent.

Literature review is another source of data in the grounded theory approach. Typically the issue remains, of how and when to engage with existing literature (Dunne 2011; Strauss & Corbin 1998, pp.49–52; Glaser & Strauss 1967, p.37). Following the traditional approach of Glaser & Strauss (1967) **chapter 6** discusses the previous findings and first deduced implications in the context of research literature. It specifies implications and gives new impulses for the emerging theory.

Chapter 7 illustrates the iteratively emerged theory, while formulating practical strategic implications for telecommunication companies conducting cross-industry innovation projects. The chapter describes a practical generic service innovation process model in detail, including sequential and parallel activities, and its embeddedness in the organizational structure. Moreover, it considers various aspects of cross-industry collaboration regarding industry specifics and it considers strategical alignment and internal transfer of the developed solution. The chapter concludes with general considerations when applying the emerged theory in practice.

Based on the practical strategic implications of the previous chapter it becomes apparent, that existing theoretical innovation process models do not reflect the characteristics of cross-industry service innovation projects in the telecommunication industry. Consequently, **chapter 8** deduces theoretical implications in form of propositions for existing theoretical process models and visualizes them in a combined extended innovation process model.

Eventually, **chapter 9** summarizes the key findings of the present study and clarifies how they contribute to practice and current research literature. In addition, it indicates limitations and suggests further topics of research.

2 Theoretical Background

This chapter describes the main terms of the research and its underlying theoretical foundation in the context of literature review.

2.1 Product – Service and Physical Good

The term product is often used inconsistently, both in theory and practice. Usually product and good is used synonymously, without any distinction between intangible service and physical good (Segelod & Jordan 2004). Rathmell (1966) defines that all economic products lie along a goods-service continuum, between the two extremes of pure goods and pure services. Influenced by the emergence of service discussions in academic literature during that time, the author identify 13 characteristics of services, including intangible nature, standards cannot be precise and cannot be inventoried (Rathmell 1966).

Based on extensive literature review, Zeithaml et al. (1985) identify that the most frequently cited characteristics of services are intangibility (non-physical nature), heterogeneity (not standardized), inseparability (of service creation and consumption) and perishability (inability to store) (Zeithaml et al. 1985). Some authors argue that intangibility, heterogeneity, inseparability and perishability is not generalizable to all services and that they fail as criteria to adequately distinguish services from goods, because many services possess one or more of the opposite characteristics, namely, tangibility, homogeneity, separability and durability (Lovelock & Gummesson 2004; Vargo & Lusch 2004b). Accordingly, Vargo & Lusch (2004) suggest that the strategy of distinguishing goods from services should be abandoned and replaced by a strategy of understanding how they are related. Consequently, the authors offer a new definition for services: "the application of specialized competences (knowledge and skills) through deeds, processes, and performances for the benefit of another entity or the entity itself" (Vargo & Lusch 2004a).

Following similar thoughts, Howells & Tether (2004) classify services into four groups, based on different transformation processes: *services dealing mainly with goods* (such as transport and logistics), *services dealing with information* (such as call centers), *knowledge-based services*, and *services dealing with people* (such as health care) (Howells & Tether 2004). Keeping this classification in mind is important for new product development, because even though the term product covers goods and services alike, they imply very different innovation approaches. Service innovation requires other organization and principles than physical products (Chen 2011), e.g., the design of processes can be more informal for services than for goods (OECD & Eurostat 2005).

Software as an intangible asset has always suffered from the tension of being both good- as well as service-oriented and the convergence of internet and classical software products and IT services as well as new technologies like service-oriented architecture led to a new dynamic (Leimbach & Friedewald 2010; Segelod & Jordan 2004). The importance of innovation in the services sector and of the services sector's contribution to economic growth is increasingly recognized and has led to a number of studies on innovation in services (OECD & Eurostat 2005). Following that recent shift towards services (European Commission & Open Innovation Strategy and Policy Group 2011), the present study focusses on innovating non-physical products. Tangible, physical complements might be necessary to access the described services, usually provided by a supplier, however, its invention, production and supply are not part of this study.

Consequently, for the purpose of this study intangibility is the most important criteria to distinguish the analyzed partly goods-like services from physical goods, allowing cheap and easy scalability, neglecting logistics, setting-up assembly lines and replication costs – similar to the common communication services portfolio of telecommunication companies. Due to economic reason, focus of the described innovation projects is to develop a homogeneous solution that can be delivered to various customers with minor adaptations and efforts, what is a main differentiation criterion to individual customer projects (Segelod & Jordan 2004). Moreover, consumption is entirely separable from the development process what is also founded in the ability to store digital data.

2.2 From Invention to Innovation

Innovation consists of two parts: the generation of an idea or invention, and the conversion of that invention into a business or other useful application. The formula 'Innovation = Invention + Exploitation' briefly describes the relationship among the concepts (Roberts 2007).

Inventions are discoveries, novel ideas, processes, methods, objects that generally result from R&D activities. They become innovations when they are transformed into marketable products or technologies, by means of investments in complementary manufacturing, technological and marketing assets. Not all inventions turn into innovations and reach the market (Giuri et al. 2007).

While the OECD and Eurostat used a very narrow definition[1] of innovation in its second edition of the Oslo Manual in 1997 (OECD & Eurostat 1997), it widened the scope in the third edition in 2005 (OECD & Eurostat 2005) taking recent research literature into account (Johannessen et al. 2001): "An innovation is the implementation of a new or significantly improved product (good or service), or process, a new marketing method, or a new organizational method in business practices, workplace organization or external relations." The authors further specify: "A product innovation is the introduction of a good or service that is new or significantly improved with respect to its characteristics or intended uses. This includes significant improvements in technical specifications, components and materials, incorporated software, user friendliness or other functional characteristics." (OECD & Eurostat 2005)

The distinction between product and process innovation is clear in respect to goods, but it may be difficult to distinguish in respect to services, especially when production, delivery and consumption of a service occur at the same time. Hence, if the innovation includes new or significantly improved characteristics to meet an external market or user's needs, it is a product innovation. If the innovation affects the organization's production process or service operations of new elements (e.g., input materials, task specifications, work and information flow, and equipment) that are used to produce a product or render a service, it is a process innovation. And if the innovation involves significant improvements in both the characteristics of the service offered to the customer and significantly

---

[1] „Technological product and process (TPP) innovations comprise implemented technologically new products and processes and significant technological improvements in products and processes. A TPP innovation has been implemented if it has been introduced on the market (product innovation) or used within a production process (process innovation). TPP innovations involve a series of scientific, technological, organisational, financial and commercial activities." (OECD & Eurostat 1997)

improved methods, equipment and/or skills used to perform the service, it is both a product and a process innovation. (OECD & Eurostat 2005; Damanpour 1996)

Most accepted definitions of innovation focus on novelty and newness. Consequently, it is necessary to distinguish the degree of innovation radicalness further, ranging across a single continuum and encompassing aspects such as what is new, how new and new to whom (Johannessen et al. 2001)? Generally radical or breakthrough innovations are those that result in fundamental changes in the activities of an organization and involve a large departure from existing practices. While incremental or continuous innovations are those that require a lesser degree of departure from existing practices (Damanpour 1996). A related concept is the one of disruptive innovation. This type of innovation has a significant impact on a market and on the economic activity of organizations in that market. The impact may change the structure of the market, create new markets or render existing products obsolete, however, the disruptive character of an innovation might not be apparent until long after it has been introduced (Christensen & Euchner 2011; OECD & Eurostat 2005). Moreover, literature generally distinguishes between new to the organization, new to the market and new to the world innovation (OECD & Eurostat 2005).

Roberts (2007) sums up: "Technologically innovative outcomes come in many forms: incremental or radical in degree; modifications of existing entities or entirely new entities; embodied in products, processes or services; oriented toward consumer, industrial or governmental use; based on various single or multiple technologies." (Roberts 2007)

2.3 Innovation Processes and Open Innovation Approaches

Various innovation process models arose within the past decades. The cross-industry innovation approach is grounded in recent open innovation developments.

2.3.1 Generations of Innovation Process Models

Nobelius (2004) notes that the concept of innovation process generations is one way to communicate different types of approaches and to describe to some extent an evolution of innovation processes. The author concerns that most companies constitute a mixture of the generations and that corresponding time period of the generations differ depending on industry segment, demographics, company age, research intensity, legislation demands, etc. (Nobelius 2004)

Du Preez & Louw (2008) and Nobelius (2004) expand Rothwell's (1992) overview of dominant innovation process models from the 1960s to today (Du Preez & Louw 2008; Nobelius 2004; Rothwell 1992). The authors identify that the first and second generation models, typically applied until the early 1970s, are sequential linear models that explain innovations by being pushed by technology or pulled by market needs (von Hippel 1976; Quinn & Mueller 1963).

The third generation describes sequential, coupling models with a series of functionally distinct, but interacting and interdependent stages, linking various in-house functions and recognizing market needs and new technologies. Cooper (1990) presented the stage-gate-model, that applies process-management methodologies to innovation process and that typically involve from four to seven stages and gates (Cooper 1990). The model allows dividing the company specific innovation processes, from ideation to launch, in a number of stages or work stations, and quality control checkpoint gates in between the stages. This creates evaluation steps that guarantee quality throughout the project and lead to early cancelation of ideas not fulfilling the expectations. Even though some authors note that the gates should be permeable to facilitate an iterative and interactive process of experimental design and exploration (Gassmann, Enkel, et al. 2010; Lynn et al. 1996), the stage-gate process model can be easily applied in organizations, and hence, is one of the most applied process models in practice.

In contrast to the linear models, the fourth generation of innovation process models emphasize cross-functional, parallel and interactive integration of innovation activities within organizations, and also stress horizontal strategic alliances, strategic vertical relationships with suppliers and coupling with leading edge customers (Niosi 1999).

The fifth generation models, from mid-1990s onwards, focus on systems integration and networking to realize a fully integrated parallel development, including strong linkages with leading edge customers and strategic integration of primary suppliers (DeSanctis et al. 2002; Iansiti & West 1997). Characteristics are extensive communication activities with the external environment and knowledge accumulation and processing (Galanakis 2006; Blomqvist et al. 2004). Despite of its focus on integration of external input sources across the entire network, the models represent a closed innovation view, because core innovation activities take secretly place within the company's boundaries.

From the beginning of the 21st century, the sixed generation of innovation models contains open innovation models, focusing on openness and collaboration across the network, allowing access to a much larger base of ideas, knowledge and technologies. These models describe a new paradigm that allows combining internal and external paths to market (Lazzarotti & Manzini 2009) (cf. figure 4). Gassmann & Enkel (2006 and 2004) identify three core processes of the open innovation approach. The (1) *outside-in process* includes the integration of external knowledge, technologies or experts in the organization. It reflects that the locus of knowledge creation does not necessarily is equal to the locus of innovation (Enkel et al. 2009). The (2) *inside-out process* includes the external commercialization of innovation while investing in new markets and licensing technologies to increase the multiplication of solutions (Inauen & Schenker-Wicki 2012; Inauen & Schenker-Wicki 2011). And the (3) *coupled process*, that includes a cooperative innovation process with complementary partners in alliances or innovation networks, characterized by integration and externalization of knowledge to reach a joint win-win situation (Gassmann & Enkel 2006; Gassmann & Enkel 2004). The authors add that in particular these companies focus on the coupled process, which are able to increase their revenues by multiplying the sales of their solutions.

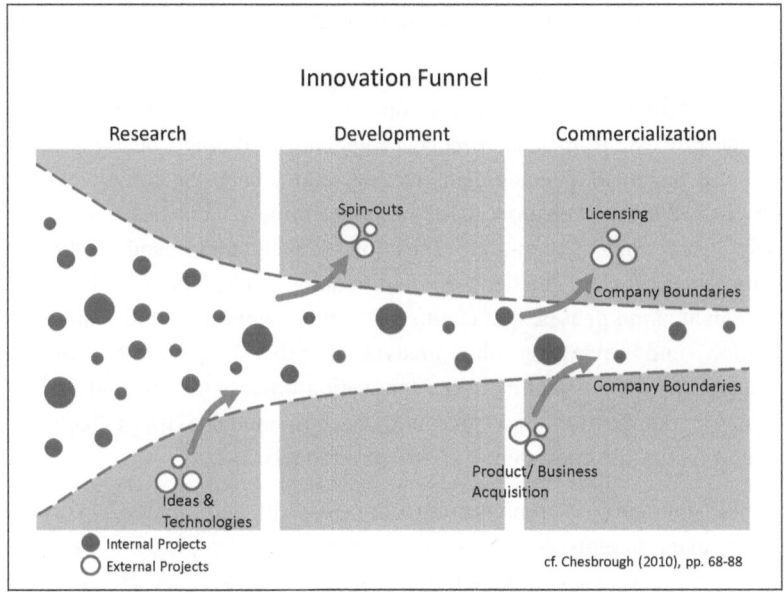

Figure 4: Innovation Funnel

2.3.2 Phases, Differentiation and Practical Integration

Models based on sequential phases differ in terms of the level of granularity. Those with just a few phases are more likely to be congruent among each other and describe real processes very well, but in a simple, abstract way. In contrast, models with many phases describe the processes of specific industries, companies or innovation types. Innovation models, hence, should be as detailed as possible and as complex as necessary.

Some authors propose an open innovation process, consisting out of three consecutive phases, namely early innovation phase / fuzzy front end, technical product development/new product development and commercialization (Gassmann, Kausch, et al. 2010). Others identify four stages of project development: idea generation, project (i.e. prototype) design, technology development and commercialization (Gales & Mansour-Cole 1995). Some authors divide the stages further, e.g.: Alam (2007) and Alam & Perry (2002) propose a ten stages service innovation model (Alam 2007; Alam & Perry 2002).

Song et al. (2009) sum up that developments in literature typically suggest a five-stage process, from idea or concept generation through business analysis, design and technical development, testing, and launch or commercialization (Song et al. 2009). Eveleens (2010) literature review identifies a very similar process from idea generation, selection, development and testing, and implementing/launch. The author additionally considers post-launch and learning/evaluation phases, that entails sustaining, supporting and scaling up the innovation and reviewing the innovation process (Eveleens 2010). In conclusion, the diversity of accepted innovation process models and its various intentions of application show that not one best model exist, rather different objectives justify the coexistence of different models.

Academic literature is representing the different requirements very well, however, even though our economy is increasingly dependent on services, today's innovation research is mainly product oriented and the services sector is still underdeveloped in terms of innovation processes (Gassmann, Enkel, et al. 2010; Thomke 2003). Most of the proposed business-to-business new service development models in literature are very similar to new product development models and generally little distinction is made if physical goods or services are applied to the processes (Song et al. 2009; Alam 2007). Opening up the service sector to the innovation process will provide new opportunities for service innovations (Gassmann, Enkel, et al. 2010; Thomke 2003). Moreover, it is getting more and more important to analyze the respective specifics, because due to an increased bundling of products and services into new industrial offerings, the processes of new product development and new service development are increasingly interdependent and cannot be conducted in isolation from one another (Kindström & Kowalkowski 2009).

The purposeful management of the innovation process is complex and involves the effective integration of *people/staffing*, planning of *organizational processes, -structures and -systems* and *plans/strategy*[2] (Yang 2007; Roberts 2007). These

---

[2]   Yang additionally suggested a 'process of quality design' that, in this paper, is considered
      to be a permanent activity in the innovation process. A fourths dimension 'systems support'
      was considered by Roberts in earlier studies, but not followed in the cited 2007 article. In

three dimensions are subject to managerial influence and need to be considered while implementing the innovation process within organizational boundaries. Roberts (2007) exemplifies the challenges that several roles of experts need to be appointed and wisely composed in order to achieve successful project results. Typically they are bound in matrix structures, namely their original discipline-based functional group and the focused project group, and hence, are influenced by competing sets of objectives. Moreover, strategic planning and strategic implementation aspects need to be considered on organizational level and technology development level. An organization's product line need to be considered while incorporating technological considerations into an overall business strategy and each stage of a technology is associated with different strategic implications (Roberts 2007). The present paper follows Roberts' thoughts. It considers the three dimensions in chapters 7.2-7.4, right after the description of the ideal practical innovation process model in chapter 7.1 and reverts to it in chapter 8.2 while discussing theoretical implications.

2.4 Cross-Industry Innovation

Cross-industry innovation is the creative imitation and retranslation of existing solutions to meet needs in other industries. Such solutions can be technologies, patents, specific knowledge, capabilities, business processes, general principles, or whole business models (Enkel & Gassmann 2010).

2.4.1 The ICT Industry and its R&D Characteristics

After various deregulation and liberalization activities, the telecommunications industry was characterized by optimism in the late 1990s, with exponential growth rate expectations for products and services (Gupta et al. 2007; Christensen & Roth 2001). The industry experienced a dynamic environment with continuous restructuring activities and an evolution from value chain to value network (Cansfield 2009; Li & Whalley 2002; Fransman 2002). Information- and communication technology organizations converged and

---

the structure of the present study 'systems support' is considered as supportive tools within the discussion of the collaboration category. (Yang 2007; Roberts 2007)

technical innovation started to challenge the companies' traditional business models. When the positive anticipations have not been fulfilled and the industry was faced with decreasing revenue streams from its communication services, the optimism changed and the industry more and more started trying to generate new revenue sources around and beyond its core-telco products and -services.

Especially internet technologies have fundamentally affected and transformed the telecommunications industry (Hess et al. 2012). Therefore, telecommunication companies developed software-specific competences, to counter complementary or substitutionary products that increasingly challenged the existing telco product portfolio (Wulf & Zarnekow 2011). Accordingly software and software-based services account for a major part of the ICT industry and are recognized as a key element in developing the information society (Leimbach & Friedewald 2010). Nowadays, all major telecommunication companies offer managed services or have IT services subsidiaries or divisions predominantly for business customers, such as T-Systems International GmbH (TSI) of Deutsche Telekom AG (DT), Vodafone Global Enterprise Inc. (VGEI) of Vodafone Group Plc. (VF), Telefónica Multinational Solutions of Telefónica, S.A. (TEF), BT Global Services plc. of British Telecom Group Plc. (BT), Orange Business Services (OBS) of Orange S.A., NTT Data Corporation of Nippon Telegraph and Telephone Corporation KK (NTT), National Computer Systems Group (NCS) of Singapore Telecommunications Ltd. (SingTel), Verizon Enterprise Solutions of Verizon Communications Inc. (VCI).

The increased diversification and the disruptive nature of technologies led to R&D activities that moved to a great extent from central research laboratories of the telecom operators to specialist equipment suppliers. Similar to other high-technology, knowledge-based industries, such as semiconductors and consumer electronics, innovation speed is high and alliances have been prevalent (Grant & Baden-Fuller 2003). The industry is characterized by a fragmented, disaggregated player landscape, that include independent, multi-country, multi-culture and multi-lingual players across the value chain/network, as well as a fragmented customer base that demands complex products and services out of a wide portfolio of technology offerings and services (Gupta et al. 2007). Consequently, the industries R&D activities are significant and involve cooperations of geographically dispersed entities and the management of globally dispersed, virtual and networked product development processes (Gupta et al. 2007). Accordingly a repository of new product development processes along the value networks exist that vary with product/service and with involved parties, and can be described as 'fairly immature and volatile' (Gupta et al. 2007; Leiponen & Drejer 2007; Li & Whalley 2002). Culture, language, perceptions, attitudes, mindsets, etc., accordingly play an important role within the innovation process that implies a loose definition and application of formalization and controlling. Consequently, players in the ICT industry should have good prerequisites to conduct innovation projects with partners from different industries as well. Hence, the industry may pioneer in conducting upcoming cross-industry innovation projects.

2.4.2 Collaboration in Innovation Projects

Substantial collaborations of two or more organizations that follow a common goal are generally described by the term 'strategic alliance'. The term implicates various forms of collaboration regarding scope and duration, and may embrace collaborating parties within an industry or cross-industries, such as supplier-buyer partnerships, joint research projects, common new product development, common distribution agreements or cross-selling arrangements (Grant & Baden-Fuller 2003). The partners' relative position classifies alliances as either vertical or horizontal. Vertical alliances are typically characterized by supplier-buyer relationships, while horizontal alliances with generally equal partners are common in projects that require relative extensive resources for an organization that are hard to raise on their own (Garrette et al. 2009).

Co-development with external partners is generally associated with higher-level innovations, specifically with more radical and more complex innovation. Collaboration helps to reduce the uncertainty inherent in the innovation process and supports the commercialization activities. Hence, collaborative innovation projects are more common when developing new to the market innovation, rather than new to the organization innovation (Kale & Singh 2009; Tether 2002). The collaborations usually involve contractual agreements or ownership affiliations as an attempt to reduce risk. However, "interfirm alliances involve cooperative relationships that are not fully defined either by formal contracts or by ownership. Hence, in terms of the theory of economic organization, they fall between the polar models of markets and hierarchies." (Grant & Baden-Fuller 2003)

3 Methodology

Guided by the research processes defined by Österle et al. (1991) and focusing the major concepts incorporated in Eisenhardt's (1989) research roadmap process, this chapter describes how the relevant theoretical concepts and procedure are applied in the present study to build practical oriented theory from case study research (Österle et al. 1991; Eisenhardt 1989). The approach is highly iterative, tightly linked to data and especially appropriate in new topic areas (Eisenhardt 1989).

3.1 General Considerations

Case study research typically describes and analyzes complex phenomena and comprises descriptive, exploratory and explanatory types of case studies (Wilde & Hess 2007; Riedl 2006; Lee et al. 1999). Research literature focusses on the descriptive and exploratory character of case studies, that account for 61 percent and 30 percent, while explanatory case studies account for 9 percent in recent literature reviews (Dubé & Paré 2003).

To judge the quality of research design, Yin (2003b) emphasizes the criteria of construct validity, internal validity, external validity and reliability: establishing correct operational measures for the concepts being studied is essential to reach *construct validity*. *Internal validity* is only a concern for explanatory/causal case studies and describes a causal relationship, whereby certain conditions are shown to lead to other conditions. Consequently, the case study should assure that the evidence collected during the data collection process is the same evidence as in the case report (Dubé & Paré 2003). *External validity* emphasizes the generalizability of the findings beyond the analyzed case study. *Reliability* allows to reach the same results when repeating the operations of a study (Yin 2003b, pp.33–39). Hence, to minimize errors and biases, the researcher should conduct the case research so that another investigator can repeat the procedures and arrive at the same conclusion. Consequently, documentation is a prerequisite containing procedures, instruments and general rules (Dubé & Paré 2003). A clear description of the data sources and the collection process increases reliability and validity of case findings (Dubé & Paré 2003). Many positivists question the trustworthiness of qualitative research and reject the concepts of validity and reliability as inappropriate criteria because they evaluate qualitative work from a positivist perspective (Cassell et al. 2006; Golafshani 2003). Advocates have responded to the raised issues and evolved the concept in recent years. The most commonly accepted construct addresses similar criteria, but uses a different terminology to distant themselves from the positivist paradigm: *confirmability* (for construct validity), *credibility* (for internal validity), *transferability* (for external validity) and *dependability* (for reliability) (Eunjung Lee et al. 2010; Shenton 2004).

Case study research includes both single- and multiple-case studies, that can be based on any mix of quantitative and qualitative evidence (Yin 2003b, pp.14–15; Eisenhardt 1989). Qualitative methods give the opportunity to gain insights into incompletely documented, new phenomena, however, concerns regarding external validity apply, even though advancements have been made over the past years (Bluhm et al. 2011; Burgelman 1983). Consequently, in many instances case studies combine quantitative and qualitative data collection methods, such as archives, in-depth interviews, questionnaires and observations, as supplements, that provide data on the same subject and will generate theory (Lee et al. 1999; Eisenhardt 1989; Yin 1981; Glaser & Strauss 1967, pp.15–18).

Qualitative research allows to generate, elaborate and test theory and recent articles focusing on theory generation have a larger impact on the accumulation of management knowledge compared to articles focusing on elaboration or testing (Bluhm et al. 2011). Moreover, various activities need to be done in parallel in qualitative research, e.g., while collecting data, the researcher simultaneously needs to think about their analytic implications (Yin 2011, pp.29–30; Idrees et al. 2011). This is why the present study follows grounded theory as central research methodology, developed by Glaser & Strauss (1967) as a systematic, comparative, iterative approach to analyze data, and to consequently, discover a theory. Grounded theory allows to inductively generate new theories while applying extensive coding and analyzing techniques, predominantly with qualitative data (Wilde & Hess 2007). A joint collecting, coding and analysis of data is the underlying operation in grounded theory, they should be done together as much as possible, and they should blur and intertwine continually until a theoretical saturation is reached (Burgelman 1983; Glaser & Strauss 1967, pp.21–43).

3.2 Selection of Cases

A conceptual framework and research questions allows to focus on the main objective and boundaries of the study (cf. chapter 1.3), while building a practical oriented theory from case study research. It supports to decide which cases to include and is the best defense against data overload in the collection and analysis process (Miles & Huberman 1994, pp.30, 55–56). Yin (2003) gives a comprehensive overview of case study research design, allowing to link the initial research question with the data to be collected and the conclusions to be drawn (Yin 2003b; Yin 2003a). Keeping the initial research question in mind, the selection of cases and the conduction of interviews is focused, but broad enough to allow serendipity during data analysis. The research question defines and guides the topic to be analyzed. Consequently, research relies on theoretical sampling not on statistical, random selection, in order to discover categories and their properties and to suggest the interrelationships into a theory (Glaser & Strauss 1967, pp.62–65). Hence, cases need to be chosen on clear rationales and not based on availability and provide clear information about the replication logic (Dubé & Paré 2003).

Cross-industry innovation is a quite new topic (as of 2011) and not widely spread, both in theory and practice. Personal background in innovation strategy and -management, and proximity to telecommunication companies is the main reason to study the cross-industry innovation process of telecommunication companies. Because of the ongoing convergence within the ICT industries, the precondition was defined, that the telecommunication company's partner company should be a company from a non-ICT industry to develop an innovative service. Moreover, the definition of the following criteria and requirements guarantee consistency and the purposeful selection of the sample.

Criteria and requirements on the cross-industry innovation projects:

- It was an innovation project, not an individual customer project, implying that the completed service was possible to sell (preferably with minor modifications) to various customers (Segelod & Jordan 2004).
- The core of the innovation project was the creation of a new service, in terms of an intangible equivalent of a good, hence, neither a personnel-intensive customer service, nor a physical product. Allowing easy

scalability of the developed solution and excluding logistics and time consuming production challenges (Segelod & Jordan 2004).

- The service was an innovation in the respective market and none of the partners had developed a similar solution before, to guarantee the innovation character and to exclude sole knowledge transfer and consulting projects.
- The innovation project was not part of an existent R&D/innovation and collaboration partnership; rather it was solely and newly planned as a project with the objective to develop a specific innovative service. Because long-term stable groups can produce rigidities as strategic partners become liable to technological conservatism, path-dependence and lock-in effects which can decrease the performance of radical innovations (Roberts 2007; Gassmann, Kausch, et al. 2010).
- The project is predominantly finished and the telecommunication company was present in all phases of the project, from ideation to abortion or commercialization, to analyze the whole innovation process from ideation to commercialization.
- Interviewees and further information material are available and academic publication in form of a case study is permitted.

Criteria and requirements on the partner companies of the telecommunication company:

- A company from a non-ICT industry, to analyze specifics of cross-industry innovation projects (Telstra Corporation Ltd. & KPMG International 2012).
- Partners were dependent on the knowledge and expertise of each other to develop the solution in an adequate time and with an adequate budget. Implying that they need partner knowledge for major parts of the intended solution.
- Major collaboration partners were up to three at least middle sized companies per project: no consumers/private persons, no start-ups, no universities or other research institutions. This restriction guarantees limited uncertainty, analyzability of structures and processes focusing on

deducing implications for telecommunication companies (Fredberg et al. 2008; Belderbos et al. 2006; Gales & Mansour-Cole 1995).

- No suppliers of existing solutions of the telecommunication company's product portfolio, because they may have incentives related to their existing products, that constrain their innovation efforts.
- No sole collaboration with partner companies IT(C)-departments, which have similar expertise, working habits, etc. across the industries, to better analyze industry specifics.

Optional, preferred and differentiation criteria:

- The telecommunication company took a leading role in project coordination and service development, because of more easy accessibility of telco representatives.
- The telecommunication company incorporated the service in its product portfolio, to gather more information about the commercialization process.
- A mutual know-how exchange and learning was necessary to develop the solution, and hence, a very intense collaboration was inevitable.
- Differentiation of possible knowledge flow directions:
  - o Predominantly from the partner company to the telecommunication company.
  - o Predominantly from the telecommunication company to the partner company.
  - o Predominantly mutual exchange.
- Differentiation of possible innovation flow directions:
  - o Predominantly from the telecommunication industry to a non-ICT industry: External knowledge helped to develop a new or to adapt an existing ICT-based service for a non-ICT industry, taking the respective industry specifics into account.
  - o Predominantly from a non-ICT industry to the telecommunication industry: External knowledge helped to develop a new or to improve an existing ICT-service in telco's core market, e.g., a service was existent in non-telco industry and either parts or the whole service got adapted to the telco industry.

    o Predominantly towards converging market: Development of a new service, based on expertise from both industries. None or only basics of the new service existed before in the telecommunication industry and in a non-ICT industry.

Intranet and internet research and first discussions with management responsibles allowed creating a project long list of 57 potential innovation projects. Unstructured interviews with project representatives, based on the defined selection criteria and the interview guideline resulted in a short list of 7 suitable projects. Case study research may cover an in-depth investigation of one case situation or it may cover in-depth investigations across multiple cases (Yin 2003b, pp.14–15; Lee et al. 1999). Multiple cases have the potential to yield more compelling evidence and are needed to develop and test more robust theories (Dubé & Paré 2003). Consequently, after further examination and considering the availability of suitable interview partners 5 projects covering 3 distinctive innovation projects define the basis for the present study. They cover success and failures, and differences regarding organizational structure, collaboration and innovation process approaches. The identification and selection of further projects was originally planned, but finally, not necessary, when during the data analysis phase it got obvious that the theoretical model reached its saturation.

3.3   Data Collection Method

Case study is a research strategy and does not imply the use of a specific data collection method (Yin 1981). The vast majority of studies applying case study research use interviews as data collection method (Dubé & Paré 2003). However, a combination of data types can be highly synergistic (Eisenhardt 1989) and increase the internal validity of the findings (Dubé & Paré 2003). Hence, a data triangulation approach was chosen, combining in-depth, semi-structured, one-to-one expert interviews, internal documents such as project plans and –reviews, intranet pages and external material such as press releases and media coverage (Golafshani 2003). The objective was to cover all phases of the new service innovation process of all selected projects, from ideation, market analysis, partner identification, design, test until the launch.

To guarantee reliable and function specific results and because usually experts did not work for a project the entire time, various interviewees were necessary for each project. Moreover, the projects typically involved both, internal roles, such as the project leader of the internal innovation project, and collaborating roles, such as the project leader of the common innovation project, and roles that were present in either internal innovation, common innovation or customer project. Discussions with members of the project steering board and project managers allowed identifying suitable interviewees from the telecommunication company and the collaborating partner companies. However, it was not possible to select the same roles as interviewees for all the projects, because they varied depending on the specific project and company. Due to the novelty of the cross-industry innovation approach Innovation Manager or NPD-Manager were not involved. The most fruitful interviews were these with Heads of Business Development, Project Owners/Heads of the respective Business Unit or Strategic Area and Project Managers.

3.4 Design of Guideline

Semi-structured interviews were chosen as main research instrument, because the research focused on a clear topic and a cross-case analysis was intended. The interview guideline iteratively evolved, from a few major questions to a comprehensive pool of mandatory and potential questions (Stigler 2005, pp.129–134).

The first guideline derived from general questions supporting the research question (cf. chapter 1.3). While evaluating potential answers, the items got classified and repeated questions got added to guarantee extensive information gathering. The questions got precisely formulated, but not too specific, to not limit the responses. The estimated response time to complete one interview should not exceed 1.5 hours. In a first iteration round with doctoral students, the measurement items got partly rephrased and expanded. Simple language was chosen to allow every project member to understand the questions, regardless of its background, job description and hierarchy. In a second iteration round with T-Systems Multimedia Solutions GmbH company representatives, the guideline was probed in a conducted pilot case and finalized based on the results (Stigler 2005, pp.129–134).

The guideline consists out of three parts and includes procedural information as a mean to increase case study reliability across the interviews and cases. The first part interrogates general information about the project, such as duration, size, organization and why it was initiated. The second part focusses on the innovation process, the collaboration with the external company and how the project differs from other innovation projects within the company. The third part deals with implications and lessons learned, as an intent to expose areas of improvement for future cross-industry innovation projects.

3.5 Realization of Data Collection

After identifying the interviewees, a first informal conversation via telephone took place, to give some background information about the research study, to explain the procedure of the interview and to schedule a meeting. The following interviews were conducted face-to-face in Bonn, Munich, Leinfelden and Friedrichshafen, Germany, predominantly in their offices or local meeting rooms, during daily working hours. In total 25 in-depth interviews have been conducted[3] and 3 of them had to be conducted via telephone. The interviews followed the prepared guideline and had a duration between 26 minutes and 1:48 hours with an arithmetic average time of 1:07 hours (median 1:05 hours). A dictating machine recorded each interview.

In the beginning it was difficult to arrange the sampling frame, because especially the managers were busy and only available after a few weeks, but during the actual data collection process, it was very easy to get the desired information and the interviewees described openly and extensively. Moreover,

---

[3] Due to confidentiality concern of one interviewee, the information and notes of only 24 interviews have been further processed and incorporated in the present study.

they generally provided project related internal documents, such as project plans, final reports, press releases, etc.

Field notes should be used extensively to process upcoming thoughts, write down new insights and interpretations, and to improve validity of qualitative findings (Dubé & Paré 2003; Burgelman 1983). General notes and reflections have been summarized during and right after the interviews. Anonymized information about the interviewees have been kept for following queries and a final version of the case studies has been sent for review. Handwritten comments and annotations in PDF files and the electronic ring binder Microsoft OneNote 2010 were used predominantly to make notes during the whole data collection and analysis process. The software tool provided an easy way to gather and structure notes, drawings and screenshots, to evolve the theory.

3.6 Data Analysis

"Analyzing data is the heart of building theory from case studies, but it is both the most difficult and the least codified part of the process." (Yin 2003b, p.139; Eisenhardt 1989) A detailed description of the analytic process allows to better understand the findings. Hence, researchers must provide sufficient information, allowing the reader to follow the derivation from the initial question to the ultimate conclusion, increasing reliability and internal validity of the findings (Dubé & Paré 2003). Miles & Huberman (1994) provide a comprehensive sourcebook with techniques and methods for analyzing qualitative data focusing on within-case and cross-case exploring, describing, and explaining (Miles & Huberman 1994).

Coding is useful for data reduction and helps to link data with the emerging theory and increases validity of qualitative findings (Dubé & Paré 2003). It does not support a lot the analysis of processes or chronology within data material, but even though the innovation process is a major part of the present study, coding, nevertheless, is inevitable, because it supports the overall research and analysis regarding collaboration and organizational concern (Roberts 2007). Consequently, the 25 recorded interviews have been transcribed into 741 pages (with an arithmetic average of 29,64 pages and a median of 29 pages), to have it in an appropriate format for QDA-software programs. For coding and querying, various QDA-software programs have been considered and the functionality of NVivo 9.0, ATLAS.ti 6.2 and MAXQDA 10 have been compared in detail. ATLAS.ti 6.2 was selected, because a free of charge license agreement was available and all three programs support the needed functionality in a similar way, to organize, document, manage, code and systematically query the given data material, and hence, allow to document, support and manage the theory generation process (Gibbs et al. 2002).

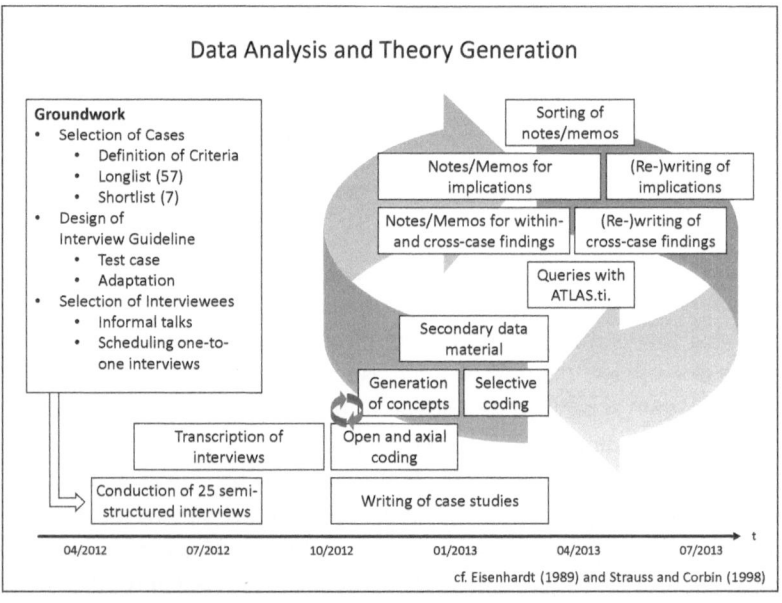

Figure 5: Chronology of Data Analysis and Theory Generation

Coding the data line by line allows the researcher to uncover new concepts and novel relationships and to systematically develop categories related to their properties and dimensions (Strauss & Corbin 1998, pp.65–71). When Glaser and Strauss independently developed their grounded theory approach of 1967, Strauss & Corbin (1998) distinguished between open-, axial- and selective coding (Heath & Cowley 2004; Strauss & Corbin 1998, p.143). In *open coding*, the analysts generate categories and their properties. The categories are systematically developed and linked in the *axial coding* process. The *selective coding* process allows to integrate and refine the categories (Strauss & Corbin

1998, p.143). After transcribing the interviews, codes evolving from the text have been applied to the document sentence-by-sentence in an initial open coding process that allowed to group collections of codes of similar context to concepts and to create a first set of concepts (cf. figure 5). In the grounded theory approach coding is more than linking codes and text fragments, it rather is a comprehensive analysis process that includes notes as preliminary fragments of the evolving theory. In an axial coding phase codes have been regrouped and the concept set has been adapted while continuing open coding in further transcripts at the same time, until a stable concept scheme evolved that allowed deriving the categories of 'market', 'project', 'collaboration' and 'company'. The names for the categories and subcategories come from the pool of concepts already discovered in data (Strauss & Corbin 1998, p.114). During further sentence-by-sentence coding of the remaining transcripts there was only need for minor adaptations of the concept scheme and quotes of the transcripts have been linked with the relevant concepts.

The following concept scheme lists the identified categories with corresponding concepts. Its description and number of relevant quotations is mentioned in parentheses:

- Market (market category consists of concepts handling information about market and industry facts)
    - o Market_Initial Situation (overall situation at the time of project initiation) (69)
    - o Market_Current Situation and Outlook (situation after realization of the project, e.g., market absorption and acceptance by the customer, economic potential, market trend/development) (34)
    - o Market_Industry Specifics (any characteristics in telco-, energy-, automotive-, health industry, e.g., regulations, knowledge, working habits) (47)
- Project (all project related concept are grouped in the project category)
    - o Project_Target and Description (objective of the project and reasons for initiation, description of the developed service) (153)
    - o Project_Scope (amount of resources such as people, man-days, budget, time) (88)

- o Project_Management (information about project management, including defined milestones, PM processes, player involvement, project structure, controlling, tracking tools, etc.) (159)
- o Project_Success (assessment by means of success criteria, e.g., technical, financial, collaboration, as well as areas of improvement) (151)
- o Project_Innovation Process Phases (chronology of innovation project and if applicable supporting customer project with a focus on its phases e.g., ideation, market/business analysis, bid for customer project, plan demonstrator, rough/detailed design, realization, launch) (196)
- Collaboration (the category collaboration groups concepts describing the mutual cross-industry work within the project)
  - o Collaboration_Condition (organization of the collaboration, intensity and frequency of collaboration, legal aspects, etc. – exclusive of collaboration tools) (62)
  - o Collaboration_Tools (all tools supporting the collaboration, such as video conferencing, data room, etc.) (50)
  - o Collaboration_Assessment (obstructive and supportive facts, such as trust, mutual reservation, common pace, privacy reason) (179)
  - o Collaboration_Telco Competences (contributed or identified superior skills and capabilities of the telecommunication companies) (41)
  - o Collaboration_Non-Telco Competences (contributed or identified superior skills and capabilities of the partner companies) (31)
- Company (all concepts concerning organizational and strategic facts within the companies boundaries are part of the company category)
  - o Company_Telco Structure and Processes (organizational characteristics and formalization of the telecommunication company, e.g., approval and allocation of resources, incentives, boards, reporting) (111)
  - o Company_Telco Strategy (overall direction and major initiatives and goals, e.g., regarding cross-industry projects, self-concept describing the desired role in the market, portfolio management, realization of speed-boat approach, etc.) (38)

o Company_Non-Telco Specifics (all aspects regarding peculiarities
of the non-telco partner companies) (34)

o Company_International and Competitor View (innovation
approaches of related groups/organizations and interdependency of
innovation with existent core business) (12)

Memos allow the researcher to keep record of the analytical process. They contain the products of coding, provide direction of the theoretical sampling and allow to sort ideas and thoughts (Strauss & Corbin 1998, pp.238–241). While coding the transcripts, code notes, theoretical notes and operational notes in form of memos and diagrams from the interviews have been continuously created to keep track of ideas related to next steps in the analysis process, the within-case analysis and deriving preliminary implications in the theory building process. At the same time further additional data such as notes of informal expert interviews, press releases and internal project documents have been analyzed and evaluated, and notes and memos about cross-case theoretical outcomes have been constantly taken to develop and adapt potential implications. Especially discussions with nine telecommunications industry experts of Australian telecommunication companies, who did not face any overlapping with the European telecommunication company and its focus markets, turned out to be fruitful in the development of implication generation and they caused serendipity effects in form of linking innovation culture aspects of start-up companies with the observed cross-industry innovation projects. Moreover, discussing first outcomes and preliminary implications with the Australian experts allowed a basic attempt of theory testing and to guarantee especially transferability and confirmability of the evolving theory to other companies and markets. At the same time a first draft of the case study write-up took on shape, that mainly was an attempt to structure the notes from the coding process. It consist of various contextual information to convey a better understanding of the 'big picture', to increase the credibility of the results and to allow to determine whether they are generalizable (Dubé & Paré 2003).

In a second coding cycle a selective coding has been applied to all transcripts, to identify further data for extending and shaping the categories, and hence, its emerging implications regarding process model, organizational conditions, cultural fit and portfolio expansion, with a strong focus on the 'project' category that has been identified as the core category. This stage of the analysis process allowed creating a storyline and aligning the other concepts and categories around the 'project' category, and to further structure the framework for the evolving theory (cf. figure 6). Notes in this phase predominantly have been taken in form of diagrams and graphics and written notes to refine the emerging theory. The grounded theory method does not provide a clear answer when the theory generation process should end. Data needs to be analyzed until patterns have clearly emerged and additional data does not provide any more value to the concepts (Burgelman 1983). Theoretical saturation is not one incident, it is a combination of the empirical limits of the data, the integration and density of the theory, and the analyst's theoretical sensitivity (Glaser & Strauss 1967, pp.60–62). First indications of theoretical saturation of the preliminary implications have been identified after around two-thirds of the transcribed interviews have been coded and analyzed, hence, it was decided not to proceed with theoretical sampling, conducting further interviews and case studies as originally considered. Moreover, five case studies is a good number to deduce and describe the innovation process, its boundaries and overarching requirements in cross-industry innovation projects. The selected projects show all identified characteristics: with and without development of a demonstrator, with and without a pilot customer, different innovation approaches particularly parallel and integrated, and they show success and failures.

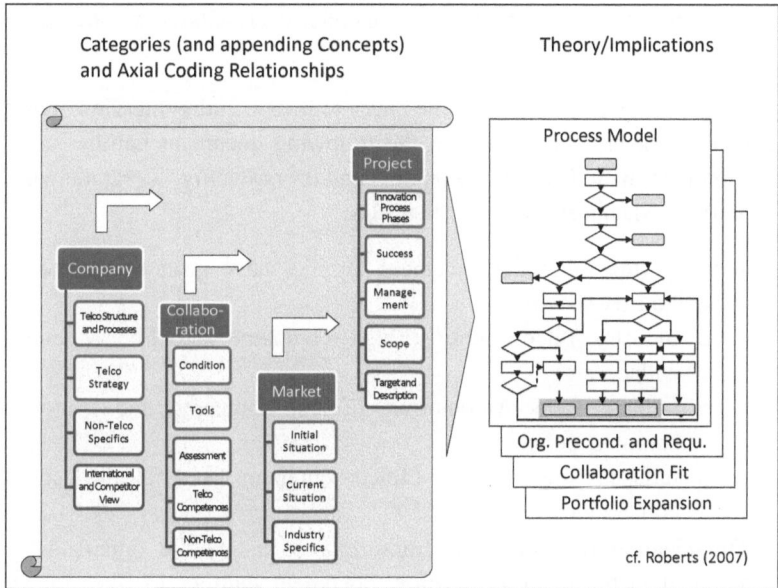

Figure 6: Categories and Concepts

In a final step all notes, memos and diagrams have been sorted and reviewed in reference to the preliminary implications, summarized in a cross-case analysis (cf. chapter 5), discussed in the context of literature review (cf. chapter 6) and have been verbally finalized (cf. chapter 7).

Queries have been conducted with the ATLAS.ti tool, both on textual and conceptual level, to reach a final theoretical saturation of the implications. Cross-case analysis, enhances the probability to capture novel findings and allows to identify patterns by looking at the data in many divergent ways, e.g., by selecting categories and concepts and identifying similarities and differences across cases (Dubé & Paré 2003). In exploratory case research a comparison of the emergent theory with conflicting and similar existing literature enhances the internal validity, generalizability, and theoretical level of the theory building from case study research (Dunne 2011; Dubé & Paré 2003; Eisenhardt 1989). To assure that the emergence of categories will not be contaminated by concepts more suited to different areas, "an effective strategy is, at first, literally to ignore the literature of theory and fact on the area under study […]". After the analytic core of categories has emerged, they can be linked to existing literature to identify similarities and convergences (Dunne 2011; Glaser & Strauss 1967, p.37).

For comprehensive retrievals with the query tool each entire interview transcript document has been coded based on the following document families, and has been combined with Boolean, semantic and/or proximity operators with the above introduced concepts and/or categories:

- Industry (the following document families have been defined to query cross-industry specifics)
  - ○ Industry_Telco-Energy (Telco company and Energy company collaborate)
  - ○ Industry_Telco-Automotive (Telco company and Automotive company collaborate)
  - ○ Industry_Telco-Health (Telco company and Health company collaborate)
- Organization (all interview transcripts of a specific organization are grouped in the mentioned below document families to filter for single companies/projects)
  - ○ Organization_Telco (Telco company is involved)
  - ○ Organization_Energy (Energy project related company is involved)
  - ○ Organization_Automotive1 (Automotive project 1 related company is involved)
  - ○ Organization_Automotive2 (Automotive project 2 related company is involved)
  - ○ Organization_Health1 (Health project 1 related organization is involved)
  - ○ Organization_Health2 (Health project 2 related organization is involved)
- Point of View (the exclusive families allow to retrieve interviews with telco or non-telco industry interviewees)
  - ○ Point of View_Telco (interviewee is employed in the telco industry)
  - ○ Point of View_Non-Telco (interviewee is not employed in the telco industry)
- Interviewee (the document families below describe in which project the interviewee has mainly worked in, to better assess individual statements)

- o Interviewee_Internal Innovation Project (Interviewee was predominantly involved in the internal innovation project)
- o Interviewee_Customer Project (interviewee was predominantly involved in the customer/pilot/demonstrator project)
- o Interviewee_Project Owner Level (interviewee cannot be assigned to either the innovation project or the customer/pilot/demonstrator project)

4 Within-Case Analysis

The following chapter is based on interviews conducted with representatives of a European telecommunication company and innovation partner companies from energy, automotive and health industry (cf. chapter 3.5). It contains three case studies, describing five cross-industry innovation projects. They follow a consistent analysis framework (Yin 1981), covering (1) Initial situation, Objectives and Description of the Service, (2) Scope and Structure, (3) Chronology and Phases of the Innovation Process, (4) Cross-Industry Collaboration and Competences, (5) Project and Company specific Organization and Processes, (6) Retrospective and Key Findings.

In a first step, all available data material have been processed and summarized to 59 pages of profound and extensive case-studies, describing the innovation projects as accurate as possible. After the final theory (cf. chapter 7) has been developed, the case studies have been condensed in a second step, forming the following revision. This step was necessary to anonymize data and to integrate only required information, to protect interviewees' personal rights and companies strategic competitive advantages.

The telecommunication company had dedicated departments and processes in place dealing with innovation management and new product development to refine and enhance its existing telco-focused product portfolio. Continuous market and technology analysis and derived business models forecasted an enormous potential for ICT related service applications across industries. As a result, the telecommunication company decided to expand its product portfolio, to increasingly approach the energy, automotive and health markets and adapted its organizational structure accordingly. The intention was to drive innovation around scalable platform solutions and to realize a speed-boat approach, with considerably degrees of freedom, that allowed developing innovative solutions without following extensive corporate processes. The mandate was explicitly specified in defining and developing services, and generating revenues. These measures were expected to help to transform the telco company and to compensate its decreasing revenues from telco services by new revenue streams generated in new markets.

4.1 Telco-Energy-Industry Service Innovation Project

Two units within the telecommunication company independently developed an identical idea of a solution for the energy industry. They joined their forces, identified the need for a partner company with energy industry specific expertise and decided to proof the initial concept under real condition. In collaboration with an utility as a pilot customer and innovation partner, and IT companies from the energy industry, the team realized a first solution for the utility and applied the insights and practical experiences to internally develop the intended solution in parallel.

(1) Initial Situation, Objectives and Description of the Service

The German energy market got liberalized in April 1998 followed by a revision in 2003. Due to the fact that no regulatory authority was formed, the following consolidation process increased the energy prices. In July 2005 the Federal Network Agency was established, to guarantee free and fair competition.

Following the awareness of on-going market liberalization activities that were similar to earlier developments in the telecommunications industry, the telecommunication company did extensive market analysis and identified that a scalable platform solution would be the basis to sell various BPO services for the energy industry. The solution also was desired by the European Union, because it replied to present demands of the energy transition, caused because more and more alternative energy got fed in the electric supply network, by private solar collectors and industrial power plants. For the telecommunication company it was possible to develop and provide the desired back-end solution and several serviced build on top. They identified an enormous need for these solutions within the following years and it would not be possible for the predominantly small- and mid-sized companies in the industry to provide these services. Existing solutions in the market had critical limitations regarding scalability, flexibility and security. Following business cases proved that the desired platform solution would provide a promising business model. EU-wide directives and national laws steadily increased the demand, what made the chance more evident for the telecommunication company to realize a successful business case.

For the telcos partner company, it was evident that German energy supplier were soon bound by law to realize various use cases. Moreover, it identified chances for managing and steering purpose, to guarantee reliable energy supply when renewable energies got more prevalent. Hence, for the utility it was important to learn technical and operational specifics of the solution and to get insights on the effect on its customers. They knew that collaborating with the telecommunication company would be an exceptional chance to gain financially supported experience in a major future topic, that would not be possible otherwise, given the small size of the utility.

The planned basic platform solution was designed to handle various further use cases as well. It was an open, scaling, multi-tenant solution that was based on an innovative architecture, that was filed for a patent. Devices from any supplier that were able to connect via standardized interface were supported. The solution ran as-a-service, based on SLAs, for electricity, water, gas and district heating services.

(2) Scope and Structure

The team of the telecommunication company was subdivided in a business development sub-team, that also provided the overall project manager, and a technical sub-team. Most of the telco-sided employees did not have any energy industry know-how. In the beginning, the team consisted out of five FTEs supported by external consultants to process industry specific information, to develop the business model including go-to-market strategies and quantity structures. The team grew to around 30 after the project proposal passed the corporate executive boards. At peak time around 60 FTE were involved, including supporting internal experts from finance, controlling, legal, sales, IT, and others.

After the team presented a project proposal to the corporate telco executive board, the innovation project was initiated. To realize the project, the team had to follow the telcos internal product development process, including various boards, to receive budget and personnel resources. The team was rejected twice in the relevant boards and it took time to pass them, because the process owners' decision criteria were not designed for highly innovative solutions. To pass the product development boards the team also had to define gates, to create a possibility to control that the project was still on track throughout the project. The investment budget for the innovation project was low double digit million Euro.

The overall project manager set an internal project management tool in place, to structure the project. Throughout the project, the team got new insights when they met with the utility, its sub-contractors or other companies from the industry. Hence, they considered following Scrum in the solution development, but because no one was experienced with agile software development they decided to stick to traditional, well-known methods.

Pilot customer project

The telco-sided team identified the need for a partner to realize the solution. It decided to conduct a pilot customer project, in parallel to the innovation project, to prove their concept and to gather and transfer industry insights for the solution development. The pilot customer project consisted out of a technical sub-team and a marketing sub-team. The technical sub-team consisted out of five to six people and got strong support by sub-contractors and the telecommunication company throughout the project. The marketing team consisted out of two people. Utility-sided one project manager was responsible for IT and processes and coordinated sub-teams. The overall project manager was provided by telco-sided business development who took the lead in overall project management, mainly because the telecommunication company provided a large part of the needed budget. He planned and conducted all meetings, coordinated the technical realization with around 22 people and organized steering-meetings, consisting out of five to six representatives of the sub-teams. All project participants were involved part-time, except the overall project manager.

Initially the project was subdivided into three phases, in addition and in parallel to the test platform development. Based on the good reputation the project generated in the energy market, a fourth phase was attached and the team realized a rollout under real conditions.

As a precondition to realize the pilot customer project and to get necessary budget releases, the team had to follow various processes and prepared internal project proposals. The proposals passed steering committee and project advisory board. Passing the project advisory board was required for all projects demanding a budget higher than a specified amount of money. The budget released was sufficient high, hence, it was not necessary to raise more budget for the first phases at a later time. The utility had to bear its own expenses, but did not have to pay for the solution either.

While setting up the partnership with the utility, a legal contract with NDA has been closed to fix the main milestones and characteristics of the solution. The project team was organized in a classical project organization and followed classical project management principles. The major milestones have been transferred to a common project plan. For major changes, the team had to pass the given processes and boards again. The procedure was just a matter of form, because the project established a strong telco executive board recognition and support, and because of declining prices for solution complementing devices.

IT-Partner

A small company supported in the initiation phase to specify the solution and provided test devices. Due to the telecommunication company's corporate standards and requirements the innovation project team had to find a bigger and more reliable partner company and at least a second sourcing company for the solution complementing devices.

One IT-Partner company was acting as a sub-contractor for the telecommunication company while supporting the development of the telco solution. The company predominantly provided the underlying business logic and industry insights, while the telecommunication company realized the major part of the development and provided the core-platform including critical customer interface. The IT-Partner company also was an IT-infrastructure supplier of the pilot customer and had expertise in industry specific and pilot customer specific processes.

(3)   Chronology and Phases of the Innovation Process

Following market analysis and ongoing EU-wide and national changes in directives, laws and regulatory policy measures, the idea to develop the service innovation evolved. It was evident that in the upcoming years it would be necessary to fulfill legal regulations and that the platform solution also would be the basis for various use cases allowing utilities to gain a competitive advantage in the increasing competition in their market. In this phase the telecommunication company got support by external consultants and telco-internal experts, especially for conducting market analysis, competitor analysis, and business model. They saw the potential to realize profitable business cases across the whole new value chain.

The telecommunication company defined the basic specifications of the solution and tried to find a supplier in the energy industry. However, all over Europe no company in the utility market was able to provide a solution with the desired specifications. Due to the fact that most players were mid-sized companies, they focused on solutions for a few utilities and were not able to fulfill the criteria of scalability, what was one of the telecommunication company's major requirements. Hence, with a strong focus on scalability and security, the initial project team designed an innovative architecture, filed it for patent and identified a need for a partner with distinct energy industry expertise to implement industry specifics and to realize the solution. They developed the technical concept further and drew up a first business model with revenue potential and strategic positioning in the market. The whole project concept was presented to the telcos executive board that gave the order to shape the business case and officially launched the innovation project. Within the following months the team extended and finalized the group business model, reassessed revenue potential, strategic positioning, etc. and included market segmentation and 10 years planning. Afterwards the executive board gave the order to realize the innovation project.

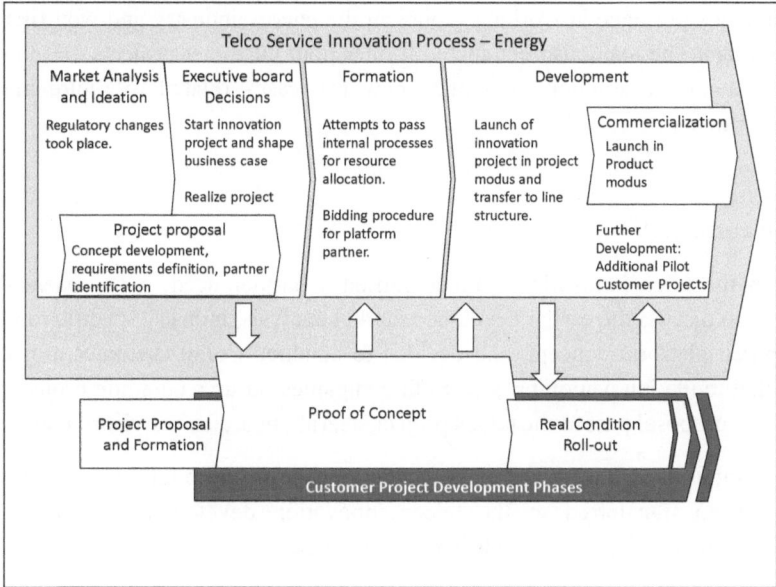

Figure 7: Innovation Project Process - Energy

The team conducted a bidding phase, based on specifications deduced from a rough concept and selected a platform realization partner (cf. figure 7). Moreover, the team was trying to pass the relevant product development processes of the telecommunication company, to allocate budget and resources to realize the solution. The project proposal was rejected twice in the boards, and hence, it was not possible to conduct the project as desired by the executive board. As a result of an internal group audit, that analyzed why an innovative project with telco executive board support was not able to pass required processes, the criteria for the processes got adapted and fast-tracks got established. As a consequence, the project passed the product development boards and got budget and resource releases, to develop the intended solution and to implement it as a new portfolio element. The team conducted the innovation project in a project mode and organizationally got transferred within the telco company.

When the solution reached maturity, first commercialization activities started in addition to the continuous platform development. Telecommunication company representatives were present at most congresses and tried to extend the company's presence in working groups in the energy industry and won first big customers. The team also conducted further pilot customer projects, to improve and extend the solution, to realize new use cases related to future market requirements.

Pilot customer project

When the telco innovation team identified a partner need, it approached the utility to discuss the earlier developed market analysis, industry specifics and the prepared platform concept, and decided to conduct a pilot customer project in parallel to the innovation project. Both companies set up a common project plan and the proposal passed the telcos project steering board and advisory board.

All insights from the proof of the concept phase in the pilot customer project have been transferred to the telcos innovation development to adapt the architecture and implement industry specific processes.

Due to the success of the project within the early phases, the telecommunication company decided to extend the pilot customer project, to realize a quick go-to-market and to incorporate more experiences into the innovation solution. To extend the project, the team had to pass internal processes again, what was possible without any delay. Hence, after the early phases the team switched from the pilot customer test platform to the desired telco live system. At that time the solution did just provide the core functionality of the desired final solution. After the integration of following insights into the platform, the pilot customer project run in operating phase to predominantly realize quality improvements based on statistical measures.

(4)   Cross-Industry Collaboration and Competences

When the telcos project team was pooled, they shared all earlier results and experiences, acted as one common team and were motivated to reach the common goal. The results of their market analysis and concept drafts have been discussed with the pilot customer and the scope of functionality of the pilot customer project has been agreed jointly. The main collaboration with the utility was within the development phase, to transfer know-how towards the platform development team. The pilot customer specified its functional demands on the solution, provided detailed knowledge and gave regular non-ICT industry process input. With the partner it was possible for the telecommunication company to develop its solution effectively and efficiently.

The collaboration with the utility was described as transparent, intensive and very good, because both partners started open minded and did not have to fear future competition. The high intensity of the collaboration was defined by individuals and got support by organizational structures, that demanded intensive communication with the project partners. Technical and marketing sub-teams met regularly, while the topics changed from fundamental industry specific questions, to assessments and definition of new features. In the beginning the companies were facing language difficulties and the telecommunication company had to learn the industry specific nomenclature, concepts, regulations, norms and business processes. It was reviewed that the telecommunication company learned a lot from the utility, especially about industry specific processes that were completely different from the processes used in the telecommunications industry and needed to be followed when acting as a service provider in the market. The utility tolerated this situation and was willing to explain because it was obvious from the beginning that the company took the role of an innovation partner beside the role of a pilot customer and it would benefit a lot from the successful realization of the project, with a minimum of invested capital and resources.

Throughout the project the utility had limited capacity, because the project was on top of its regular work. And also the team members of the telco sometimes needed to prioritize daily work higher than the common project. This situation was accepted by both partners, however the utility in some situations wished to get more information from the telco.

E-mail, telephone and presence meetings have been the predominant forms of communication. In the beginning telephone conferencing was not familiar to some utility employees because it was not part of their daily business and not anchored in their culture, but it was quickly established as a common communication medium. It was used for weekly meetings within the technical sub-team, the rest of the team preferred presence meetings and e-mail. A telco owned data room solution was used for common collaboration, but the team considered various other solutions, because at that time the internal transfer price was unreasonable high. Even though most telco people had to travel far to the pilot customer, telephone conference was not considered as necessary and the team predominantly scheduled meetings at the utilities site.

The telecommunication company had specific competences. Especially security topics related to data transportation and data processing were underrepresented in the utility industry. The telcos approaches were considered as too extensive in the beginning, but got quickly accepted, also because EU-guidelines picked up the topic as well. The utility provided various information regarding energy industry specifics and made it possible for the telecommunication company to develop its service innovation solution. Processes and ways of working differed a lot. The utility as a public service did not have an innovation process because it only needed to react on legislative changes what was very seldom.

When the telco-sided team got the innovation budget release through the telcos product development process, for the innovation project, it decided independently from the utility the scope of functionality, depending also on the knowledge acquired within the pilot customer project collaboration. When the team replaced the test platform at the customer with the desired service innovation platform, the collaboration with the utility was more like a customer-supplier relationship than the earlier innovation partnership with strong knowledge exchange. The utility felt uncomfortable with the new telco solution and described it as a black-box, because they did not know anything about the internal processes and were not able to reconstruct measures.

(5)   Project and Company specific Organization and Processes

When the telco executive board accepted the internal business case for the innovation project and gave the order to realize the service platform idea, the team had to pass the telecommunication company's product development process. Within this process the team had to apply for budget and resource allocation in the respective boards and prepared the relevant documents including business case. The project was rejected twice in the boards' prioritization rounds, because it did not reach the minimum revenue numbers. Hence, the team did not even get the chance to present the idea in the boards. The project had to compete with other incremental innovation products related to the telecommunication company's existing portfolio that had much higher short term revenue expectations. The team stimulated various decision makers within the telecommunication company, including the internal group audit, to figure out how to get a budget release and to realize the project as desired. The internal group audit came to the conclusion that mainly the prioritization criteria were not innovation friendly. As a result the telcos product development process had been adapted for innovation projects and the revenues got assessed in three years estimation and not in a decade outlook. Due to the changed condition and the increased top-management attention and support, the team was able to pass the required boards.

Following the resource releases from the internal product development process, the project organization persisted, but the team was transferred within the organization. As a consequence various people in the line needed reliable information and binding statements from the team mangers, e.g., the product management demanded a new function and the team had to decide if and when that function got part of the solution. They got more accountable, because their decisions and acting had a direct impact on the telecommunication company's revenues. With the transfer in the line organization, the team lost a bit of its degree of freedom to act independent from other stake holders.

Based on the development of the customer test platform, the team got industry insights, proofed its developed concept and adapted required specifications for the telcos service innovation platform. The team was bound in steering and reporting processes of the entity that financed the pilot customer project. When the telco-sided team applied for the budget release through the telecommunication company's product development process, it developed the customer test platform and the service innovation platform in parallel and accordingly had to follow two process environments. These processes created a high workload for the responsible project managers. The team also needed to follow the processes, because it was dependent on supporting entities such as procurement, legal, etc. The number of contact people for the team to deal with increased with the maturity of the solution. This was in accordance with the processes for the day-to-day business, but not appropriate for time-critical innovation projects. In the end the project team was able to get just one contact person per entity, what made it much easier for the project team and led to faster support.

(6)   Retrospective and Key Findings

It took around one year for the team to pass the telco internal new product development process and it only was possible to pass it, because guided by the top-management the selection criteria for innovation projects have been adapted. After passing the process, the support of the top-management also helped to get suitable experts in the project and to make it easier for the team to convince other managers in the group to collaborate. However, the board attention also caused that various internal stakeholders from strategy, staff positions, controlling, marketing and others demanded information and reporting. Hence, stakeholder management was a heavy workload for the team. The utility did not notice difficulties with the internal innovation processes, because the innovation project were detached, conducted in parallel and had to follow completely different processes.

The utility and the IT-partners accepted the leading role of the telecommunication company, what resulted in some restraints regarding their desired value chain, because acting under the financially well-funded telco umbrella still made the project a success for them. Also, some telco-sided limitations in the course of balancing line activities and project responsibilities, and especially a lack of communication while implementing the final solution at the customer-site, led to irritations, but got accepted. Beside the financial facts, it also got accepted, because the partners also were facing limitations. Moreover, the utility was able to generate a massive nationwide publicity and was one of the first energy suppliers in Germany who got extensive knowledge in the specific field. In the end, the conduction of a customer project in parallel to the innovation project turned out to be a good approach. It enabled the telco-sided team to proof the designed concept and to gather expansive industry insights that were incorporated in the development of the service platform.

When the service innovation project reached maturity, the solution was integrated as-a-service in the telecommunication company's product portfolio and successfully launched in the market. The solution was superior to other new solutions in the market, ran successful and the telecommunication company won other big customers.

Assessed by technical measures, the solution was a success, even though some features were not yet subjoined. For the telecommunication company the project fell behind the financial numbers mentioned in the business cases, but it was assessed that decreased revenue growth was based on delays in regulatory directives and acts, and that as a consequence the planned revenues were able to reach at a later point of time. There was no doubt that beside the technical success, the solution would succeed in terms of financial indicators as well. Various following customer project proved the assessment and even during the project more and more utilities realized the advantages of this kind of BPO. Moreover, for the telecommunication company the solution was the technical basis for further use cases.

4.2 Telco-Automotive-Industry Service Innovation Project

The telecommunication company decided to develop a scaling white-label
solution for the automotive industry, offered as-a-service. To attract pilot
customers for the planned solution, the telecommunication company and its
innovation partner company developed a demonstrator as a first and major
instance. It turned out that the market accepted the scope of functions of the
planned solution, but demanded a proprietary approach and followed a multi-
vendor strategy. While ongoing commercialization efforts, the partners
considered modified approaches of the solution. At the same time, expansive
strategical and organizational changes affected the telecommunication company.
As a consequence, a re-assessment of all automotive related innovation activities
took place and the company decided to focus on other innovation projects that
allowed a more sustainable way to act in the new market.

(1) Initial Situation, Objectives and Description of the Service

The telecommunication company analyzed trends and markets, and identified
car-ICT as an important converging field. It was anticipated that within a few
years 70 to 80 percent of all 10 to 12 million newly sold vehicles in Germany
got equipped with GSM-, UMTS- and LTE-based connectivity. Since
connectivity was the telcos core business, the company wanted to participate at
the predicted revenues. The company also wanted to avoid a probable price fight
with its competitors selling ordinary M2M modules, and hence, planned to
develop and operate a more sophisticated solution, what allowed connecting
vehicles with its environment, via a back-end system. The solution should give
its users the ability to use various predominantly internet-based services while
driving, without driver distraction and should allow vehicle manufacturers,
rental companies, etc. to increase its customer service and manage the car more
efficiently.

The idea itself evolved from earlier projects in logistics and road charge, where
similar proprietary solutions have been set in place, but limited to a specific set
of services and not flexible and stable enough to fulfill the scaling requirements
of automobile manufacturers.

An automotive supplier company that was a customer of the telecommunication company was notably affected by the fact that more and more functionality moved from within the vehicle towards the cloud and the hardware business constantly changed towards a commodity business. The company did not have the same level of in-house expertise in the field of car-ICT and back-end infrastructure compared to its main competitors, and hence, needed a partner to develop the market and to acquire expertise in this field. The automotive supplier identified the telecommunication company as a reliable, well-known partner who was able to complement its abilities and who was willing itself to act in the converging market.

None of the two companies was able to accomplish the project on its own, with a reasonable amount of resources and time, and they determined that its competences complemented each other. Moreover, the automotive supplier saw the chance to integrate the telecommunication company's expertise, while the telco had a distinct interest in intensifying its relationship with the automotive supplier, because the company was one of its major customers. Hence, both companies decided to realize synergies, to move forward and defined the solution in detail.

First innovation project

The telco-internal business model described various commercialization options including revenue sharing model and considered the back-end platform usage as a basis component in for following customer projects. The internally calculated business case was attractive for the telecommunication company, financially, to strengthen its relationship with the automotive supplier, to gain experience for further projects in the automotive market and in terms of reputation.

The first step towards the planned solution was to set up a prototype to showcase its functionality, because based on its industry knowledge the innovation partner expressed that there would be no vehicle manufacturer willing to order the solution without testing and seeing it. In a second step, the solution itself should be developed in collaboration with a pilot customer. The telecommunication company agreed in higher pre-investments to build a demonstrator, and hesitated to develop the solution without a pilot customer even though this approach was also common in the automotive industry, when it was foreseeable that the solution would meet the potential customers' expectations to a specific degree. Hence, both companies decided to develop a demonstrator as a first instance, to attract a pilot customer. It was planned to include all major functionality the later solution should have, but no non-functional criteria such as scalability and security.

The fact that the innovation partner company was a tier-one-supplier for potential customers, one of the telcos biggest customers and itself saw a market potential in commercializing the solution, made it evident, that the go-to-market for the new service would primarily be led by the partner company. The telecommunication company aimed at operating the infrastructure as-a-service, offered as a white-label solution. After various commercialization activities following the public presentation of the demonstrator, both partners realized that it was hardly possible to sell the intended solution, as visualized in the demonstrator, hence to serve other automotive related use cases, the team discussed evolutions of the service.

Second innovation project

Similar to the first solution, the second one promised the chance for the telecommunication company to develop a platform for the automotive market, to strengthen its relationship with the innovation partner, and to achieve a good reputation in the evolving market. There was no need to develop a demonstrator and the partner company already acquired a pilot customer.

The second innovation project aimed at supporting the development of a mobility solution, focusing on providing information about electric vehicles to the user or owner, to manage the vehicle, and to allow a power supplier managing the charging process. The solution was not based on the earlier developed demonstrator, and the originally considered platform was designed but not sufficiently developed. Since system stability was required and there was not enough time to develop a new platform, the team decided to expand an existing matured platform, which was developed earlier by the telecommunication company.

The innovation partner provided a required device and the telco provided mobile communication interface, to connect the vehicle to the back-end application.

When offering the solution to another potential customer, it would be possible to replace the players by its respective competitors, but given the fact that both partners already did investments and proved expertise in the specific solution, they had an advantage compared to its competitors.

(2)   Scope and Structure

At the beginning of the collaboration both companies set a steering committee in place to mutually discuss all major activities for the common innovation project and to better solve potential internal, cross-entity challenges and prioritization of resources. It consisted out of five to seven responsibles and met regularly. The managers also involved the top-management to attract attention and support, to more easily solve potential barriers.

First innovation project

In the ideation and assessment phase the team consisted predominantly out of five to ten solution and industry experts, analyzing the feasibility of various different solutions. Different telco-internal business cases have been designed for various potential projects in the automotive market.

When both companies started its collaboration activities, a joint activity in a non-exclusive partnership was agreed, fixed in NDA, MoU and a short partnership contract, regulating exploitation rights of the solution and the activities until the first milestone, the presentation of the demonstrator. It was agreed that both companies bear the expenses for their own parts, which should be roughly equal parts. A common business model was designed and a common project schedule with milestones has jointly been set in place and its progress was tracked with MS Project. They did not share a common project office but in the end phase, just before the presentation of the demonstrator, the technical experts physically collaborated.

Telco-sided, various entities have been involved, that were physically spread and like in other projects as well, the telco followed internal project management guidelines for its project management, resource planning, creating specifications and conducting the development. Both companies employed around 10 people.

The project was cross-financed by various telco entities, what forced the team to follow steering processes of these entities for the approval of innovation budget and to prepare different gate templates. A first pre-budget was released to analyze the feasibility, based on pragmatically designed telco-internal business cases and a later initial funding was provided for realization. In total both companies invested a low single digit million Euro.

Second innovation project

Given the contractual situation, the partner company was overall responsible in the face of the pilot customer and the telecommunication company was a sub-contractor, but in the project collaboration both partners were equal. A common project schedule has jointly been set in place, with milestones and all ideas of both partners have been considered and discussed. The partnership was non-exclusive and regulated by NDA and a partnering contract.

The partner company took the project management lead and organized common telcos, steering committees and administrated a project plan with MS Project, containing a time schedule but no resource plan. The project management did not follow a predefined, standardized innovation process, because the partner company's processes were too extensive and did not consider cross-industry partner involvement. Moreover, the team was small and composed the solution based on existent standard components.

Both companies provided one project manager for the project, to steer the activities autarkic towards its own organization. Telco-sided 10 to 15 software developers in the development phase, and two delivery FTE in the end phase were involved. Partner company-sided three FTE were involved and steered the partners' sub-teams and some sub-contractors.

In the initial phase budget was released by different telco-entities. Further budget was refused, when the solution reached basic functionality. As a reason resources got bound in other projects and just were sporadically accessible to realize the scope desired by the team. Innovation partner-sided the budget was slightly higher and not exhausted. The company was willing to invest further and to develop further use cases.

(3)   Chronology and Phases of the Innovation Process

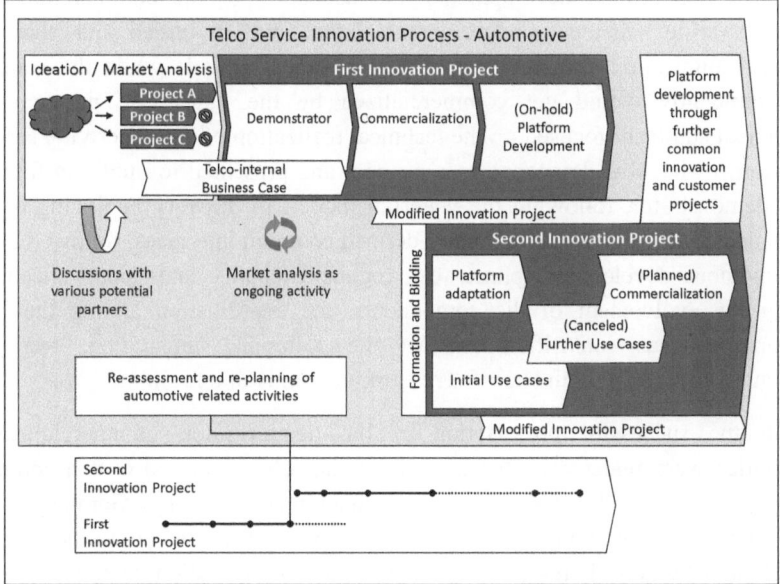

Figure 8: Innovation Project Process - Automotive

First innovation project

In the early ideation, screening and assessment phase, internal telco-entities collaborated intensively. A team around industry and solution experts created a short list of three potential new innovation projects. One of them was later continued as first innovation project (cf. figure 8). The team received a very limited pre-budget, to continue with further evaluation and to develop a pragmatic business case. Possible platform designs have also been rudimentarily assessed at that time. The decision to move the car-ICT topic forward and the identified need of an industry partner led to discussions with various potential partners, and it was decided to intensify the collaboration with one innovation partner company. Early talks also included the common definition of use cases and how to realize them. After an early budget release, one part of the telco-team focused on developing the common and internal business cases, while in parallel another sub-team focused on the technical realization. The business case team also designed a common business model with the partner company, and later on iteratively further developed the telco-internal self-sustaining business case, compliant to existing steering processes and taking the increased maturity of the solution and market developments into account. For the commercialization phase various methods were contemplated, but it turned out that the telecommunication company would take a supplying role while the solution predominantly would get commercialized by the innovation partner. The technical sub-team focused on the technical realization of the underlying service platform, what was developed as a concept and presented to public in form of the demonstrator, following the waterfall model. In the very beginning of the technical realization both companies defined common interfaces, followed by an autonomous development phase with regular exchange and update meetings. With the evolvement of the components, the coordination among the both companies was intensified and in the following integration, test and reconciliation phase both teams were working jointly at one site.

With the successful presentation of the demonstrator, commercialization activities were underway. Based on following discussions also at a common roadshow to all major vehicle manufacturers in Germany, the telecommunication company internally integrated advancements to its platform concept. However, it turned out that the commercialization of the originally planned solution was more difficult than expected. In the discussions with the vehicle manufacturers it was evident, that they did not want an all-in-one solution, but rather desired modules to have the flexibility to choose each component from different competing suppliers. Moreover, the vehicle manufacturer considered the solution as part of their area of influence and were hardly willing to buy a white-label platform solution. Accordingly the telecommunication company knew that it was not possible to reach the initially calculated revenues, the earlier planned business case had to be re-assessed and the solution had to be adapted to market it. The telecommunication company focused on developing the platform within a first pilot customer project. It was evident that it was possible to develop the infrastructure in collaboration with the partner company, however, with a strong focus on openness and flexibility to easily integrate other hardware suppliers as well. In addition, and as a result of the good collaboration, both companies followed international commercialization paths for the demonstrator and considered modified approaches for their target markets.

Second innovation project

When a manager of the innovation partner company identified a pilot customer for an electro mobility solution, he contacted the telecommunication company for potential collaboration. He did not screen the market for potential other partners. In-depth market research and market analysis have not been conducted by the telecommunication company for the specific project, because they had confidence in the innovation partner who assessed the electro mobility market in general as very profitable.

For the bidding phase of the electro mobility project the telco-sided team needed to prepare project proposals and business cases, and had to pass various gates to get internal budget approval for the adaptation of an earlier developed platform and the future development of different use cases. Various entities of the telecommunication company provided resources and innovation budget. The bidding procedure did result in an order for the innovation partner, and both partners moved the topic forward with the innovation partner as the general contractor and the telecommunication company as a sub-contractor. The partners brainstormed several use cases, and after an assessment and prioritization procedure, where both partners had to balance their opinions regarding the importance of potential services, the team defined use cases and reassessed the budget needed. For the use cases the telco-sided team had to prepare project proposals and business cases, and had to pass various gates to get budget releases, very similar to the processes followed for the bidding phase. Within the use case development phase the telecommunication company was hardly able to revert to components earlier developed in the first innovation project but used an earlier developed platform as a basis and adapted it to the project's needs. The development and pilot phase contained overlapping activities in the beginning, of development, test and adaptation of use cases and adaptation of the platform.

After a stable version in the pilot customer project was reached and presented to public, the telecommunication company did not plan further financial support to improve the solution and to include further use cases. Hence, an originally planned phase covering a reassessment of use cases and the planning of activities for the next year was canceled and the development of earlier defined use cases just sporadically continued.

(4) Cross-Industry Collaboration and Competences

When both companies decided to collaborate in a common innovation project they initially set up a common steering committee. The committee met every week in the beginning of the first innovation project and changed to every two weeks while the second innovation project was in its pilot phase, to discuss common goals and approaches, both via telephone and physically. For the partners it was a new situation to collaborate in an innovation project with an external partner and it took some time and openness to understand the respective foci mainly based on different product life cycles, business models and accountability in the industries. Balancing the different cultures, laws and working habits of the industries was a new challenge for both companies. The main competences of both partner companies completed each other. For each of the partners it would not have been possible to build the end-to-end solution on its own.

First innovation project

In the early phase of the first innovation project the respective sub-teams set up a common business case and had regular calls of two hours per week and in total three physical meetings to align the common business model. Both partners only shared information that were necessary for reaching the common goal. In this phase the team responsible for the business model predominantly used e-mail as the main communication tool. It was reviewed that a state-of-the-art collaboration environment would have been an advantage. The team responsible for the technical realization of the first innovation project used more sophisticated collaboration tools and met regularly, physical and via telephone conferences. The innovation partner provided a WebEx-solution and the telecommunication company a common development and integration server. The server was also used to exchange data packages, but did not provide the functional range of state-of-the-art collaboration solutions. The technical realization teams had their distinct working packages corresponding to their competences and were discussing various tasks mutually. They collaborated intensively and tightly while specifying the solutions interfaces in the beginning. In the following realization phase was little need to collaborate because both parties focused on the development of their working packages. In the test and migration phase the collaboration intensified, the technical telco sub-team was working at a innovation partner site to harmonize the developed components, what was described as very beneficial to adjust the components and to reach a cultural fit in that extend that company borders blurred and the two teams literally became one. Overall the collaboration was described as very good, cooperative and pleasant, with faithful relationships between individuals. People of both companies showed a strong commitment, were willing to collaborate and were motivated to finish the showcase. After the demonstrator was presented, and first common collaboration activities undertaken, the collaboration was less intensive for the technical teams. Predominantly because the technical development was on-hold and both partners recognized that the demanded model of a module based solution implicated that the partner might work in the next order with the direct competitor. Hence, both partners were more sensible in sharing knowledge because the fear increased that the shared knowledge reaches straight its competitors. Nevertheless, the commercialization was a common task and wherever possible both companies presented the solution together, even though the innovation partner had better contacts to potential customers and given the fact that generally vehicle manufacturers initiated a bidding procedure for the single components. For a sales call to Asia the team used a video conferencing solution that was part of the telecommunication company's product portfolio. The main reason was to demonstrate the innovation partner the solution and to potentially sell it.

Second innovation project

In the initialization of the second innovation project the team did not have a common nomenclature and people used different definitions for various terms related to the new topic of electro mobility. This situation was temporarily and did not hinder the collaboration in a sustainable way. The core management and steering committee that collaborated since the earlier first innovation project did not identify the need for further reconciliation measures and specific criteria for the collaboration have not been defined. Both companies jointly determined responsibilities within the project, assessed the budget, potential external expertise, created a project plan, signed a non-exclusive partnership contract and NDA, and answered the call for proposals. The contract design phase was described as rather demanding because topics such as privacy, liability, etc. had to be discussed.

When the project realization started, a common brainstorming was a joint activity to define and prioritize use cases, based on market insights. Both partners partly had different opinions regarding the needed services and implementation of the solution, and it took some time to establish a mutual understanding for the working habit of the partner. In the end the partners agreed to develop some initial use cases and to define further use cases at a later time. The back-end-sided development and adaptation was solely done by the telecommunication company. The team reverted to an earlier developed platform that already provided most of the needed functionality. Hence, the team's predominant task was to adapt the system with reasonable resources and to develop and align interfaces in collaboration with the innovation partner to connect the back-end infrastructure with vehicle components.

In all phases the project teams of both companies collaborated and communicated regularly to inform the partner about respective progress using mainly e-mail, telephone, regularly telcos twice a week, WebEx and presence meetings, but no data rooms. All documents have been sent via e-mail what led to extensive data traffic. In retrospect a data room was identified by a telco manager as a meaningful complement. For tracking and tracing of change requests and error messages, the innovation partner provided a web-based open source project management tool on its server.

Both companies' competences complemented each other, they accepted each other as partners and communicated it externally. The collaboration was described as very close, very open and was considered mutually as very good on the personal level. However, the collaboration got worse when telco-sided the budget got limited, what affected the development progress. Moreover, it was communicated relatively late to the innovation partner, that there was no budget anymore for the realization of further use cases.

The pilot customer got the solution for a very competitive price, was not interested and involved in the realization, gave the project team plenty of scope and allowed the project team to define the scope of services during the project, and accepted all the proposed use cases. The collaboration with the customer was described as very cooperative.

(5) Project and Company specific Organization and Processes

The project idea emerged within two different telco entities as part of their regular market analysis process and as an innovative approach to intensify its relationship with one of its major customers, to stimulate its revenues. Both intentions based on their internal incentive structure. One telco entity was not used to invest big amounts of money because it predominantly conducted customer projects, while for the other, as well as for the innovation partner, it was more common.

After the telecommunication company conducted a major organizational restructuring, it re-assessed all strategic activities for the automotive market and decided to minimize its investments for the first and second innovation projects, because of strategic and financial reasons. For telco-internal political reasons it was not able to cease the projects immediately, because of internal interdependency on company-wide personnel resources to conduct innovation projects, and hence it was necessary to balance various interests. Moreover, the fact that the innovation partner was one of the telecommunication company's major customers deferred its decision as well.

In the beginning of the first innovation project the team followed classical processes, which usually handled the identification of a customers need, design and provision of an architecture, conducted by company-wide expert teams. For budget releases the telecommunication company had to follow internal processes of two entities that partly did not run in line. Hence, it was challenging but necessary for the project team to embed the project processes into the company's ones, to make use of the available company budget and personnel resources. But nevertheless, dedicated experts were hard to allocate and were ,partly bound in other tasks.

To start the second innovation project the telco-team needed a budget release to pay internal resources. It had to prepare a business case and a project proposal and had to pass various telco-specific and entity-specific gates. The budget was calculated to adapt the back-end infrastructure and to develop initial use cases. When the project team wanted to extend the earlier defined use cases in coordination with the partner company, the released budget was not sufficient and boards had to be passed again. After the reorganization, the telecommunication company decreased its investments significantly, and thus, it was nearly impossible for the team to realize additional use cases beyond the minimal promised solution. The innovation partner team did not face similar budget restraints from within its company. The necessary budget was assessed and released in the beginning, people assigned and due to the fact that there were no processes in place for cross-industry innovation projects, the team was able to act relatively detached from the company's structure and processes but was strongly influenced by the given innovation processes.

(6) Retrospective and Key Findings

The first innovation project was initiated by two telco-entities in collaboration with its innovation partner. When organizational restructuring took place, the project got assigned to a new entity. When the new entity re-assessed the telcos automotive activities, it decided to focus on other innovation projects because it did not believe in a sustainable business model and identified a strategic risk to act tightly with the innovation partner. The fact that the new entity had monetary goals compared to innovation departments, made it more difficult for the team members to get support for the project. While one entity aimed to increase short-term revenues in customer projects, the other entity's focus was to develop strategic topics including scaling network solutions (cf. figure 9). Moreover, the new entity had a bigger concern that the innovation partner itself wanted to acquire infrastructure knowledge and to act as a competitor in the market. Collaboration inevitable led to an exchange of knowledge, and the surplus was higher for the innovation partner. Moreover, although both partners were officially equal in the projects, the innovation partner had a more powerful position in the collaboration, because they were a major customer for various other telco projects and as a supplier for vehicle manufacturers they had better contacts to potential customers.

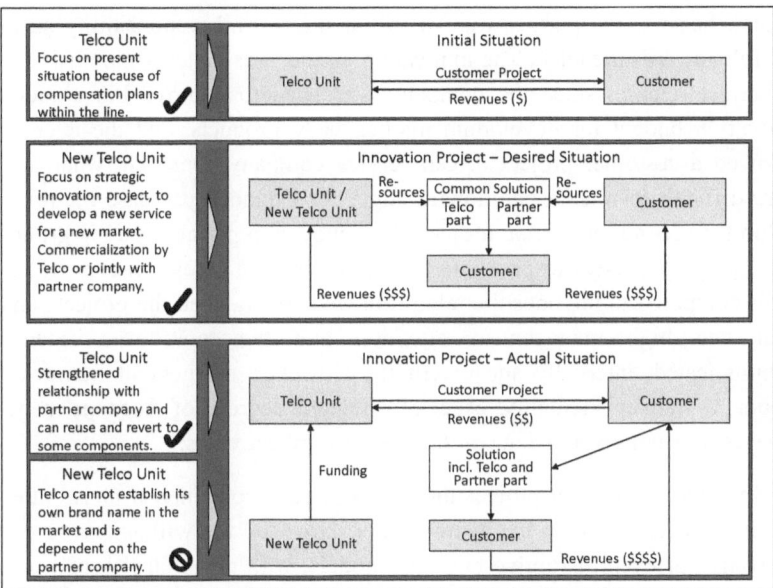

Figure 9: Intentions of Innovation Project - Automotive

Telco-internal predefined approval processes caused a considerably amount of workload for the telco-sided project team, especially for budget releases and resource allocations. On the other hand, quality gates were less present compared to traditional innovation projects. In the first innovation project the team had to pass various boards and when the budget was released in the early stage, the technical telco project team was facing the challenge to get the right people on board, because many resources were bound in other projects and were not easily to replace. In the end of the second innovation project the budget was decreased significantly what made it almost impossible to proceed. Telco-sided budget overall was a scarce resource in the second innovation project, while the initialization and especially at the end. Privacy was a topic not considered in the beginning of the second innovation project, but quickly came up and the team needed professional support. Hence, boards had to be passed again for budget and resource approval. In general, topics related with privacy, intellectual property and contracts were difficult to handle for the team both, within the telecommunication company and with the innovation partner. Within the telco because the group had extensive processes in place and the team was facing extensive structures and processes, and in collaboration with the innovation partner, because the partner wanted to put the overall responsibility as far as possible towards the telco. The innovation partner was better positioned in terms of budget release, since the company was used to approve big amounts of innovation budget for developing market-ready products, and the telco entity followed a customer project driven service company approach, what made it more difficult to release additional budget for extended use case requirements within the second innovation project. The innovation partner as a big group also had various processes to pass that took time, but the whole process was much faster compared to the ones the telco followed. Telco-sided the projects and the team had high management attention and the topic was extensively communicated, internally and externally. However, the general framework to support a start-up mentality, especially related degrees of freedom in budget releases, was not given, what complicated several activities.

For the first innovation project the business case contained a phase after the successful completion of the demonstrator, however, the willingness to discuss innovative or even disruptive business models, with potential risk and revenue sharing aspects, was underrepresented. Both parties stayed close to its core products and desired a non-exclusive partnership, mainly because operative executive manager with short-term KPI incentives formed the steering board and because both companies aimed at increasing their part of the value chain. The desired high innovation capability demanded a medium-term perspective and the willingness to implement expansive changes, what did not correlate with the managers short term revenue goals. This rather conservative attitude on both sides led to the fact that, e.g., scaling business models were not further discussed.

Various aspects need to be considered while judging the success of the innovation projects. In both projects, the telecommunication company did not develop a platform, as a new portfolio element, as originally intended. In the first innovation project the telecommunication company aimed at increasing market acceptance in developing the solution within a customer project, what was not possible because no vehicle manufacturer agreed to serve as a pilot customer. Hence, the platform was designed but not developed. The second innovation project based on an aged architecture that was stable, shortly available but that was not possible to serve as a common basis solution providing commonalities for several customer needs. Moreover, the underlying business case was not attractive enough for the new designed entity that had overall strategic concerns, and hence, decided not to proceed beyond an ordinary customer project. In addition, both innovation projects did not generate a positive financial result. In the first innovation project the predicted turnover deployed in the business case was not reached, because it was not possible to commercialize the demonstrated solution as predicted, and the second innovation project had predominantly non-financial objectives. However, both projects contributed indirectly, because the development of the demonstrator was accomplished successfully and the electro mobility project provided a pre-agreed customer solution. Both strengthened the partnership with the innovation partner, one of the telcos most valuable customers and led to further customer projects. The team got deeper insights and expertise in the automotive and electro mobility market, what resulted in more innovative ideas that were not related to the original ideas. And most important it was possible to position the telecommunication company as an innovation- and project partner for following projects in the new industries.

4.3   Telco-Health-Industry Service Innovation Project

The main components of the described service solution have been developed within two innovation projects. Developed predominantly in collaboration with a partner company and supported by hospitals, the developed solution forms the nucleus of the telecommunication company's health platform, that was considered to provide various service solutions for the health industry. The solution for heart failure was the first use case of the developed service solution and the first solution on the telco's health platform.

(1)   Initial Situation, Objectives and Description of the Service

Heart failure was one of the most common causes of death in Germany and caused immense high treatment costs. By the means of the developed solution, it was expected to cut costs, especially in structurally weak areas where cardiologists are not available within a reasonable radius, and it was strived for increasing the life quality of the patients, because they did not have to go to the hospital so frequently and had a subjective higher feeling of security, because of their daily adjudged vital signs. However, at that time, no appropriate solution was available on the market.

To further accomplish the transformation of the telecommunication company, to tackle increased pressure in its traditional market, the company decided to develop new services for the health industry and to commercialize the services in the market. In this regard, and after extensive market research that proposed a big market potential for a scalable service approach, the main task was to develop a platform as a basis for various health services related solutions.

The telecommunication company had health industry experience through various previous customer projects, but decided not to develop the platform internally, following the closed innovation approach, because they still were dependent on the knowledge and especially industry expertise of partners. Moreover, it realized that a collaboration approach would increase the possibility of meeting the market needs.

The telecommunication company saw the opportunity to consolidate its knowledge from previous health projects in the context of the first innovation project, to develop the core components for its desired health platform in a customer project and with the support of a partner from the health industry. The telecommunication company was aware of the risk of developing a health solution, since the market was in the formation phase, competent partners were difficult to find, and high investment costs were necessary. Moreover, the health industry was characterized by a high degree of complexity, because many players were involved, e.g., the developed solution helped hospitals to lower their costs and increased care, but they only were able to use the solution if health insurance funds accepted to pay the treatment. Hence, lobbying played an important role and a proximity to politics was beneficial. The telecommunication company did not act visible in the market until the first innovation project, but slowly built up recognition value.

The Health Platform

The health platform was considered as the basis platform for various health related solutions for diseases, such as heart failure, diabetes mellitus, COPD, mucoviscidosis, high risk pregnancy, etc. It was planned to steadily develop the platform, to integrate corresponding modules for further disease pattern and to offer the solutions as-a-service.

The solution for heart failure was the nucleus of the platform and provided the first basic modules such as access and authentication, workflow handling, processing, prioritization services, etc. All basis services and modules have been developed in the first innovation project. The following innovation project based on the developed components and added further modules to the platform, based on customer requested features. With increasing maturity it was possible to completely revert to the developed modules integrated in the platform and to minimize manual adaptation work.

The solution by itself was a software based solution, running on server, what allowed offering various operation possibilities, either at the customer site or as-a-service. Given the German legal situation, codified in the Medicinal Devices Act, a declaration of conformity was a prerequisite to launch the service. Moreover, the solution had to comply with Service Level Agreements determining the availability of the system, because the solution was designed for high risk patients and any downtimes and failures could have fatal consequences.

First innovation project

A structurally weak area was suffering a lack of medical service provisioning and provided aid money to launch an innovative solution. The scope was not just to build up one solution for each of the hospitals, but rather an integration of hospitals in one system environment and in one solution.

The telecommunication company saw the potential to start the development of its planned health platform through a customer project and to develop the first modules, while its partner company wanted to stimulate the demand for its medical devices. For the hospitals the solution was a chance to reduce costs and to realize competitive advantages. After successfully passing a bidding procedure, the consortium developed the workplace, server infrastructure and devices at the point-of-care for the solution, serving high-risk patients with heart failure.

Due to data security concern the hospitals decided to operate the solution in their own server infrastructure. The next step with regard to the telecommunication company's scalable platform approach was to develop a multi-client capable version operated within the telecommunication company's data center and to offer the solution as-a-service.

Second innovation project

The second innovation project was a large-scaled clinical trial aiming to improve the care of heart failure patients in structurally weak areas by the means of an innovative service solution. The participation in this trial allowed the telecommunication company to add features and modules to its platform, developed in the first innovation project. The major enhancement was to host components of the solution in the data center of the telecommunication company pioneering for the first as-a-service customer. By the end of the second innovation project, the telecommunication company was prepared to offer the solution as-a-service in a multi-client enabled environment.

(2) Scope and Structure

First innovation project

The telecommunication company discussed the possibility to undertake the first innovation project as a general contractor. Due to market constellations, desires and specifications the telecommunication company and its innovation partner jointly accomplished the project in a consortium constellation. The project team telco-sided decided to grow constantly and gradually, due to the requirement: 'Think big, start small.' In the initial phase of the first innovation project the team consisted out of five to six people. New thematic constellations and ranges of issues requested additional experts. The team grew to around fifteen to twenty people. In the end phase of project around fifty people were involved, in various teams such as deployment, service, support, etc. Some of the system components have been developed by international spread telco entities. The team worked tight with its innovation partner and was in consultation with other partners. The innovation project budget was low single digit million Euro aid money, and all expenditures within the bidding phase had to be raised by the applying companies. The telecommunication company provided a budget for the demonstrator version, that identified the needs of the medical professional world, went further than existing solutions and found general approval. Hence, the joint bit of the telecommunication company and its partner company got accepted and further budget was internally authorized, because the aid money budget would have been sufficient for a traditional customer project, but not for the development of the core components of the telecommunication company's health platform.

Second innovation project

Various organizations have been involved in in the trial, which included various medical use cases. The telecommunication company collaborated closely with its innovation partner and the hospital representatives. In the development phase of the innovation project around 35 telco developers have been involved, and innovation partner-sided further 8 to 10 people, partly as FTE and part-time. Public aid money was provided, but it was expected that the companies contribute.

The telecommunication company took the lead for the technical development and guided its innovation partner, in both innovation projects. The telco set up a rough time master plan and aligned it with the project attached organizations. Following the telcos common practice of project management, codified in internal project management guidelines, this plan also contained a detailed technical planning. The basic approach was to develop the service solution and the telcos health platform respectively based on the typically used waterfall model. Due to the size and diversity of desired use cases, the team changed to agile software development. For the first innovation project the requirement demands existed in form of the demonstrator, developed during the bidding phase. Based on the demonstrator the team derived development documentation and specification, data model, etc. For the second innovation project requirements have been defined by the participating hospitals. Hence, some of the requirements only made sense in the context of the trial, but others were relevant for sophisticated service solutions and got integrated in the telecommunication company's health platform.

(3) Chronology and Phases of the Innovation Process

Various health related customer projects have been conducted by the telecommunication company before the innovation projects started, and corresponding expertise has been acquired. As a continuous task, academic paper and market analysis have been reviewed, market studies and business cases have been performed and possible technical solutions have been discussed with technical experts and sales to guarantee a smooth place of the topic in the market (cf. figure 10). The majority of the analysis attested a lucrative market potential for the development of a health platform, and the developed service solution in particular. Hence, the telecommunication company decided to develop a solution accordingly, in case a respective customer request arises. The requirement of a customer project was set to increase the congruence with market needs.

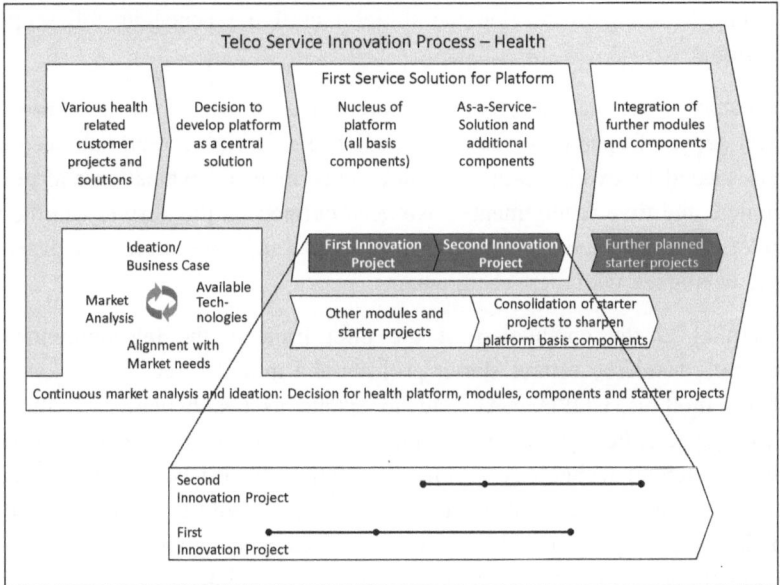

Figure 10: Innovation Project Process - Health

The successful accomplishment of a clinical trial led to an allotment of aid money to build up the desired service solution. Two hospitals brought the first innovation project in being with a request for information, what was the start of a defined and structured bidding phase. The telecommunication company saw the chance to realize its platform approach within a pilot customer project, identified suitable partner companies, hold exploratory discussions and defined its collaboration with a suitable innovation partner. As a first instance the partners gathered requirements and defined the service self-financed, with the hospitals in various rounds of negotiation. They developed a demonstrator to visualize and showcase the hospitals demands in their proposed solution, and adapted it steadily. Various telco-boards had to be passed and altogether around a quarter to one third was part of preparing processes, followed by conception and specification phases. They started the project realization following agile software development and delivered the first components. The development continued and it was possible to complete additional components and enroll first patients on the platform. It was also possible to integrate first experiences from the second innovation project as well as experiences from the first customers regarding handling to develop the solution until it reached its full scope of operation.

When starting the second innovation project, a large part of the first innovation project scope of operation was developed. The existing solution had been used in the second innovation project requirements analysis, evaluation and design. Technical and time requirements have been defined by the hospitals in the role of pilot customers and the telecommunication company's project management had to be aligned with these requirements.

In parallel to the conduction of the pilot projects the telecommunication company considered further starter projects and modules such as solutions for diabetes mellitus, COPD, AAL, etc. to extend its health platform within the following years. For this reason the telco was in contact with medical experts as well as potential technology partner companies. It also consolidated existing projects to sharpen the platforms basis components. With the completion of the two innovation projects, the telecommunication company had reached a functional range that prepared it to offer the solution as-a-service to other customers.

(4)   Cross-Industry Collaboration and Competences

The telecommunication company was able to acquire comprehensive know-how in data processing and data aggregation in earlier innovation projects and committed the same team for the present service solution. It defined and specified interface standards and interfaces for the transmission of vital signs. Based on the specifications the technical partner company realized the interfaces and provided market expertise, political penetration and the point-of-tier through its medical devices. The partner company was a vehicle for the telco to make the first move in the field of the developed service solution and to act as a supplier in the market. The hospitals in the first innovation project took the role of customers and provided necessary requirements and background information. In the second innovation project one hospital was in the role of development partner and customer who defined the requirements. The hospital had outstanding medical knowledge related to the developed service solution, because of a previous clinical trial, and it had international visibility and reputation in this field. Without the medical input of the hospital the project would not have been possible. the telecommunication company's competence was the system integration, in detail the integration and assembling of various modules, the development of diverse services, as well as its medical know-how from various previous projects, and additionally know-how in project management, project coordination, project steering, operation of the solution in the telco's own data centers and in particular the integration of communication services. The hospitals and innovationpartner provided detailed health industry related expertise.

First innovation project

In the initial phase of the first innovation project, the technical partner also collaborated with a competitor of the telecommunication company, but quickly ceased it to fully focus on the cooperation. After both companies decided the basic conditions of their collaboration during the bidding phase of the first innovation project, LOI and NDA have been signed. In further negotiations detailed contracts regulated the rights each party claimed on specific parts of the development. No success criteria have been defined for the collaboration neither within the consortium nor with the external partners.

Various collaboration tools have been applied in the first innovation project for internal and external collaboration. Presence meeting was the most common collaboration method with the hospitals, especially in the bidding phase where delicate topics have been discussed and to get to know each other. Later on e-mail has been used more intensive and access to dedicated rooms in cloud collaboration environments have been provided. Videoconferencing was unusual for the hospitals, but got quickly accepted. The telecommunication company worked with internationally distributed software development teams. The collaboration with these teams as well as with its partner company predominantly took place via e-mail, telephone and cloud collaboration environments. Fax, chat and messaging have not been used and public available solutions such as Skype were not allowed due to group security and guidelines.

Communication with the hospitals and with the partner company was essential for the success of the first innovation project, because it strengthened the inter-personal relationships, what was the basis for a good collaboration. The collaboration was described as good and cooperative on each level and the partner made effort to harmonize the development processes. However, the partner company expected the telecommunication company to intensify its sales and marketing activity to promote the common solution. In general, the collaboration was characterized by phases of euphoria and demotivation, because of the highly challenging technical requirements. The final retrospective judgment showed that the collaboration was successful also on an inter-personal level and that there was the will for further cooperation.

Second innovation project

In the second innovation project the collaboration with the partner company and internal entities directly resulted from the collaboration in the first innovation project and respective communication processes were partly well-rehearsed. There were no mutual reservation and no reason to reduce the collaboration. The hospital took the role of a customer who defined the requirements and the role of a development partner at the same time. Characteristics applied in the communication with nursing staff that were not used to design requirements documents suitable for software development. The communication was most intense in the beginning, when the requirements got discussed. Success criteria were defined by the research questions of the clinical trial, namely to define statements about treatments with the new service solutions compared to traditional cardiologic care, and secondly to prove the superiority of the developed service solution in structurally weak areas. To fulfill the trial and to get significant results, technical criteria had to be reached, such as availability of the system, response time, reactivation in case of failure, to be on schedule while developing the solution components and to host and operate the solution for the hospital.

Irregular presence meetings have been conducted with all involved organizations. The technical consortium used internal weekly telephone jour fixes, organized by the telecommunication company's project manager and on the internal operational level mainly presence meetings, but also video conferencing and cloud collaboration tools for development. Video conferencing was a mean to cut travel expenses, however the availability of technically equipped rooms was a constraint. In general, the team made the experience that a combination of teleconference and presence meetings was the most effective and efficient way of collaboration. The telco project manager reported regularly to the service solution responsible in the telecommunication company who reported regularly to the head of the health unit.

(5) Project and Company specific Organization and Processes

The telecommunication company team did innovation projects not as a daily business, it usually worked customer project centric in the context of software development projects. In this case the project team expanded the scope of the project and designed a solution that was able to be reused in various use-cases. The accomplishment of the innovation project was highly dependent on the involved group of people and not based on standardized innovation project regulation, due to the fact that a general master map did not exist for cross-industry innovation projects. Potentials had to be analyzed individually and respective approaches identified. The initiative and the engagement was not external given, but came from within the group. This inevitably led to increased expenses and issues with involved departments and managers, because the project team had to operate within the given structure, culture and processes – e.g., employees were assigned to the line organization managers who had fiscal goals, and the internal reporting and decision chain reached up to the group board. The team had to follow strict reporting processes and had to pass various boards. Also, they needed various approvals from the group privacy and big obstacles were staffing and procurement- and ordering processes. All these processes were time consuming and cost personnel resources, because forms had to be filled, people contacted, fixed dates met and information compiled. At the same time the team had to focus the accomplishment of its mission, namely to finish the innovation project with adequate time, budget and quality.

The telecommunication company's structured and standardized process approach, internally slowed down and complicated some activities. The smaller innovation partner considered this approach as new and was willing to revert to the telco's knowledge and experience and was pleased to make use of it within the project, while the hospital marginally noticed that the processes led to delays, but hence, saw room for improvement.

(6) Retrospective and Key Findings

When the telecommunication company identified potential partners to realize the first innovation project, they decided not to move forward with some companies, because they feared a drain of knowledge, leading to increased competition in the new market. The telco found a potential innovative partner that complemented its innovation approach, and hence, sounded collaboration potential. The telco was dependent on the collaboration with a partner from the health industry to develop its health platform, especially in the design and development phase, because it suffered a lack of health specific knowledge and contacts within the industry, which were very important in the highly regulated market environment. The dependency on a partner decreased with the maturity of the solution.

Due to the fact that the telecommunication company decided to accomplish the health platform development by means of customer projects, staff related synergies appeared, e.g., the whole team involved in the development was composed, incorporated and well-rehearsed, and the project manager of the platform innovation project was the project manager of the first innovation project at the same time. For the second innovation project another project manager was chosen due to the partly time overlapping project schedule and the given workload. The approach to solely innovate with a pilot customer mandate was judged by the people of the telecommunication company as superior compared to an approach where a solution is developed without an expressed customer request, because of an increased focus on market needs.

The telecommunication company's standard processes were not laid out for radical innovation projects, resulting in additional efforts for organizational measures. The company's guidelines and processes were evident in all phases of the project, e.g., while ordering a server, or assigning employees with specific know-how, inevitably let to delays in the schedule.

Especially in the second innovation project it turned out to be a new task for medical staff to prepare a requirements document for software development and to conduct a requirements analysis, mainly because of the different professional nomenclature and working habits. It was a learning process, not just to bring together two different world of thinking, but also to reconsider expressed requirements. It was important to use the same vocabulary to talk about the same issues. That was a process of harmonization what was extremely important to identify the correct requirements and to develop inter-personal relationships. The personal relationship, trust and transparent communication with the partner and customer was judged as essential for the success of the project and as much more important than just regular status e-mails or official jour-fixes.

By the end of the first innovation project, the telecommunication company was able to market a basic service solution for heart failure and was trying to attract further customers. The realization of both innovation projects was followed by various inquiries from potential customers, to develop adjacent solutions. This approved that the telco's solution met the market needs and that the company was able to establish its brand in the market.

5 Cross-Case Analysis

This chapter outlines findings across all projects. It is an intermediate step to the generated theory and structures notes evolved from the case studies.

5.1 Market and Industry Specifics and Commonalities

Across all innovation projects the companies' reason for participating in the innovation projects was because ICT was affecting their core markets and they expected a direct or indirect financial benefit from the collaboration. They aimed at increasing the sales of their existing products, building up a future business or partnership with the telecommunication company, or quickly acquiring expertise in a predominantly ICT dominated new field to realize a financially attractive business case with an innovative solution, all as part of their companies strategy or because they followed legal regulations. Likewise, in the ICT industry, for the telecommunication company it seemed promising to acquire non-ICT industry specific expertise to develop and offer innovative B2B(2C) solutions with distinctive ICT functionality in the respective non-ICT market.

It was pointed out several times, that by pooling the own knowledge and competences with the industry specific knowledge and competences of the partner company it was possible to develop the solution in a much shorter time and with much less resources compared to a non-cross-industry innovation approach. However, across all projects it also was expressed that literally two worlds encountered in the projects, in terms of industry specifics such as product innovation/life cycles, innovation speed and commercialization approaches. Compared to the ICT industry, where every few months a new product got launched, and hence, the innovation speed was quite fast, in the energy industries it took up to 20 years to replace, e.g., specific components of the energy project, and hence, the innovation speed was much slower. This also influenced the attitude towards error rate a lot. In the ICT industry updating, bug-fixing, roll-back and golden-disc-procedures were common, but in the Health, Energy and Automotive industries a completely different mind-set was prevalent and the companies desired to invest much more time in testing and traceable development to exclude possible errors as much as possible. Moreover, they wanted to assure, to meet either legal regulations or requirements of a few dominating customers.

Across all industries, organizations followed a multi-vendor sourcing strategy to reduce risk by the dependency on one supplier and they hesitated to accept a complete solution as a form of BPO. They rather desired to initiate bidding processes for individual solution components and to operate a solution on their own instead of as-a-service, because they feared to lose control over a valuable asset. Economical valuation changed their perception over the years.

The number of players in the specific industry areas was limited and the experts knew each other. They regularly met on congresses, in committees and in federations. Hence, an extensive network and the participation in lobbying activities were beneficial to improve its visibility in the respective markets. The telecommunication company slowly built up its presence and was recognized in the industries rather as providing an enabling role, than as a new competitor for existing players.

Market acceptance of the developed solutions varied a lot across the projects. The energy solution and the original automotive solution followed a B2B2C approach. The energy solution was accepted and demanded by the business customers, because the utilities were forced by legal regulation and the new service was a welcomed form of BPO to lower their costs. However, the majority of the end customers were not interested in paying a premium for a limited increase in functionality. The original automotive project was not accepted by the business customers, because of the manufacturers' strategic reasons; however, the solution was accepted and desired by the end customers. The automotive partner company decided not to approach the end-customers directly, because at that time it was against its commercialization strategy. The second automotive solution and the health solution followed the B2B approach. Both were accepted by the respective business customers, but the commercialization took long, because the automotive project served a new trend and potential customers were scarce, and due to the German health market specifics the health solution needed health funds to pay the care.

5.2 Innovation Project Characteristics

Interdependency between differences and similarities of the innovation processes, with its structural peculiarities early got obvious while analyzing the innovation projects.

5.2.1 Structure and Management

The case studies visualize two approaches of conducting an innovation project: with a pilot customer project and without a pilot customer project. The former was the desired approach, because it reduced risk, both financially and in terms of meeting market needs. The pilot customer paid a little amount for its desired solution, and hence, the needed pre-invest was limited. Moreover, an innovation that was already proven within a customer project increased the commercialization potential a lot. Consequently, the innovation teams were trying to win pilot customers, defined the premise and designed the solution to conduct the innovation project as soon as a pilot customer was present.

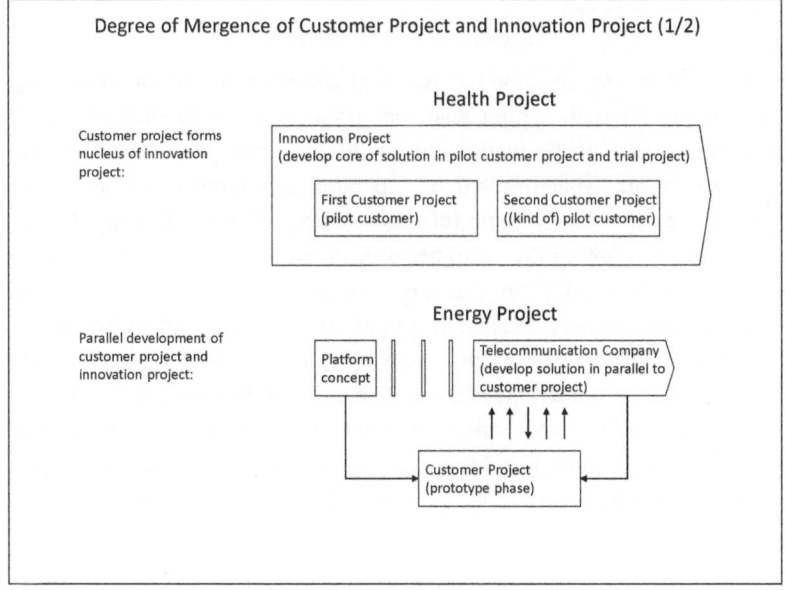

Figure 11: Project Structures (1/2)

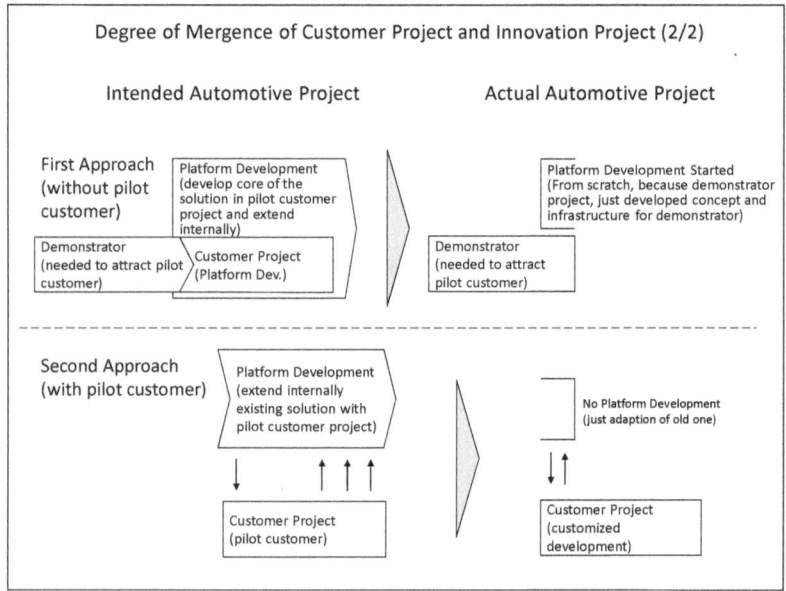

Figure 12: Project Structures (2/2)

The case studies also described a structural characteristic when conducting the innovation project with a pilot customer project: The customer project either formed the nucleus of the innovation, or the customer project was a means to prove the concept, what resulted in a parallel development of an innovation solution and a customer solution (cf. figure 11 and 12). Developing the core of the desired innovation in the customer project was the cheapest and fastest way to conduct the innovation project, however, it extended the time to finish the customer project. Mainly because it was necessary to develop components in a more extended way to focus on a scalable platform, serving general market needs, not single customer specifics, and because of the increased complexity, it was necessary to put more effort in a sustainable development (Segelod & Jordan 2004) (cf. figure 13). Moreover, for a new service, legal and privacy assessments, product descriptions, etc. had to be prepared as well.

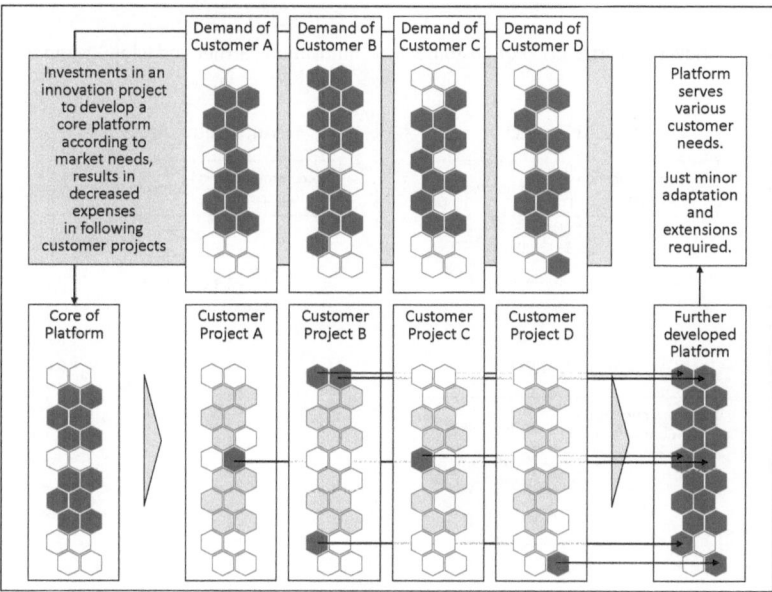

Figure 13: Platform Innovation Project

It also got obvious that the innovation partner was able to fulfill the role of a pilot customer at the same time, or different organizations fulfilled these roles (cf. figure 14). The integration of a pilot customer into the innovation process or the intensity of its supporting role varied across the projects. The pilot customer either just defined its requirements or collaborated to a much greater extend, depending on the existence of one or more other partner companies, the degree of innovativeness of the solution and its dependency on industry specifics.

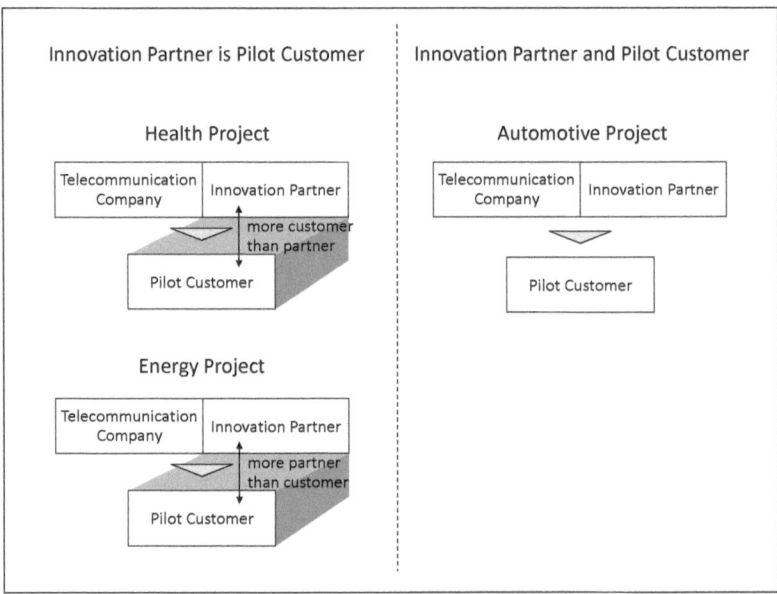

Figure 14: Innovation Partner and Pilot Customer

All innovation and customer projects followed classical project management models, including among others, project plan with defined milestones, project manager and steering board. The innovation project teams either conducted the customer projects in addition to the innovation project, or they steered it and appointed a team that conducted it. Regardless the structural organization of the projects, each project was managed individually, the representatives of the partner companies were only involved in the project management of the project they dealt with and each project had to pass respective boards. Generally the projects were less formalized, including less project controlling and tracking, compared with traditional daily business projects, but nevertheless, it was assessed that the telecommunication company-sided internal workload on management-level was determined by more than 50 percent by communication with internal stakeholders. As soon as the projects were conducted within the boundaries of the organizationally exposed strategic areas, the situation only improved a little bit, because the strategic areas were still dependent on other entities across the telecommunication company group.

Predominantly in software development the coordination of internationally distributed sub-teams was common. Difficulties in coordinating activities sometimes appeared for the telecommunication company-sided project managers, but also occasionally at the side of the partner companies. The traditionally waterfall model or V-model usually used for software development projects was too inflexible for innovation projects. Agile software development turned out to be more appropriate for innovation projects due to the possibility to adjust the requirements within the development process and its output speed. However, a lack of experience with this development approach partly hindered the adoption.

5.2.2 Process Phases

All innovation projects passed through various equal stages, such as ideation, market analysis, design, development, test and commercialization. Across the projects the stages have partially been conducted in parallel and with varying degree of severity and level of granularity.

Getting aware of new business potentials in non-ICT industries was a result of earlier customer projects and on-going market analysis. Ideation and market analysis were parallel activities and influenced each other. When a team decided to examine a specific topic in more detail, the assessment included a business component with first business case and a technical component with rough design of a concept. This was necessary to raise innovation budget to continue in project mode. Within this elaboration phase the make-or-buy decision was made and a need for a non-ICT industry partner was identified. The decision for a specific partner was influenced by market screenings and the respective project environment.

Various internal organizational and structural decisions and actions followed to sustainably form the internal innovation project, and partly at the same time the partnership formation with the external partner company or consortium took place. The later mainly included a more detailed design and adaptation of the technical concept and common business model, legal contracts and common project plan. From that time, insights from the collaboration with the partner influenced the internal elaborations and assessments. The role of a potential customer and its potential supportive input varied a lot across the projects and reached from merely requirements definition to involvement as innovation partner. Regardless of the role, its input also helped to shape the concept design of the planned solution in the innovation project. Even though the cross-industry team started to develop a common solution within the innovation project or within the customer project, it was evident that especially the telecommunication company's internal business case and the technical core components stayed in-house.

In the following realization phase the partners developed the solution and regularly tested it in parallel. It also was common to adapt and extend the solution, given new industry specific input. The organizational structure of the innovation project determined to what extent the telco's internal team was developing an additional internal solution based on the outcomes of the common solution. One potential approach to integrate further modules and to realize different use cases in the innovation project was to initiate further customer projects or to start first commercialization activities, both before the final state of the innovation was reached. Major new modules and new use cases have been conducted and formed as new projects, with all organizational and process-related activities, either in the old partner/customer constellation or with new partners/customers. Commercialization was either a common or a sole activity of the telecommunication company, dependent on the earlier agreed contracts.

5.3   Cross-Industry Collaboration

Across all projects the cross-industry collaboration was repeatedly metaphorically illustrated with the same statement: literally people from two worlds met each other. Based on the earlier described industry specifics, especially regarding innovation approaches, the involved people had different working habits, used different nomenclature and different matured innovation processes. Due to the fact that the people usually were open minded, willing to collaborate and willing to understand culture, behavior, processes and 'periods of thinking', and as a result of often, intensive and transparent face-to-face communication, especially in the beginning of the collaboration, it was possible to quickly establish supportive relationships characterized by confidence and trust, and to harmonize working habits. It was assessed across the projects that good relationships on the personal level were much more important than well-defined formal processes, and that personal meetings were the best means to reach this. Moreover, the employees needed a different, more flexible mind-set than their colleagues in the group.

Partnership contracts and NDAs defined the collaboration and have been designed in the beginning of the projects. Scope, depth and validity of the contracts varied a lot across the projects, as well as early discussions related to the commercialization, beyond the development phase. The collaboration intensity with the external partners increased significantly after the initial discussions about internal phases of ideation and market analysis, to define the partnership, the working packages of the solution and partially to develop a common business model. The technical working packages have been handled individually because they predominantly based on the companies competences, but communication, regarding status updates, common planning and especially know-how transfer, continuously was high until the realization phase came to an end. The collaboration intensity in the commercialization phase varied across the projects and was determined by contracts, organizational constellation, and the financially more powerful partner, who was in a better position to innovate, to extract and exploit expertise from the foreign industry and to commercialize the solution. In general, the collaboration intensity in the customer projects was higher than in the innovation projects.

Throughout the projects, the prevalent collaboration tools such as telephone, telephone conference and e-mail have been frequently used, and the technical involved members used or desired more sophisticated collaboration environments such as web meeting, secure data room and development environments. Video conferencing was not common, because the availability of technically equipped rooms was a constraint and the teams were not willing to make efforts to set-up video conferencing solutions, because alternative tools such as web meeting and telephone conference were available.

5.4 Organizational Environment

The telco's structure and processes changed a lot during the observation period, essentially to establish a more innovation friendly environment. Accordingly the innovation projects got affected by its organizational environment to a predominantly great extent.

5.4.1 Traditional Resource Allocation

All described innovation projects needed a modest budget to assess the general feasibility and profitability of an initial idea. The needed budget for internal and external resources was generally provided relatively non-bureaucratically and quickly by the closest management responsibles, group innovation budget boards and/or the business development entity. After a positive assessment a substantial innovation budget to realize the solution was needed. The budget release processes for this much bigger amount of budget and the corresponding application and reporting processes differed significantly among the projects.

In the energy project, the team followed the telecommunication company's traditional new product development process that was predominantly designed for high-volume, incremental, core-telco product related innovation. It happened twice to the energy project that the team did not pass specific boards or gates, even though the project had executive board support of the telecommunication company. Hence, the team had to generate alternative budget sources across the group to get support and to maintain at least a basic operation mode. The telco's internal group auditing department investigated why it was so difficult to conduct innovation projects and identified among others that in the traditional product development processes, the estimated sales volume was one of the key criteria that lead to the cancellation of disruptive innovation projects, because the sales figures of these projects were generally smaller compared to incremental innovations of present products. One manager of the energy project recalled the project assessment in the internal group audit report: "A small, very committed acting group of people within [the company], has managed to develop a product concept against all resistance and while bypassing present processes [...]". He entitled the energy project as a submarine within the telecommunication company until its product development process got adapted to support non-core-telco related innovation with a mid-term business case. As a consequence of the subsequent budget release, the team was able to staff the project with experts and increased its operational activities significantly.

The automotive project was cross-financed by the telco's innovation entity, its B2B-Sales and its B2B-board's innovation budget. It was conducted within the B2B entity's boundaries, and hence, the team had no point of contact with the telecommunication company's product innovation process. This made it easier to launch the project, however, the released amount of budget was not as high as necessary to develop a complete solution. The automotive company generally conducted innovation projects and produced components with internal budgets prior of having a specific customer. This approach was not common for the telco's B2B entity, that was a customer project driven company and generally conducted projects according to specific customer orders, innovation projects took a negligible role in its business. Hence, telco's B2B entity was more limited regarding innovation budget releases and faced internal structural barriers when a released budget needed to be increased. The investment for a demonstrator was a challenge to justify and a budget release to develop a complete solution was not possible to allocate, even though the investment might have led to a more promising positioning in commercializing the solution in the market. Moreover, managers of the telco's B2B entity focused on sending its experts into customer projects rather than in the innovation project, because it was in line with their personal incentive model. Hence, the staffing took longer and partly necessary experts were not available.

5.4.2 Organizational and Procedural Changes took place

The telecommunication company identified serious deficits while conducting non-core-telco service related innovation projects, among others in the group audit report that assessed that classical processes were not suitable to drive these kinds of projects. As a consequence, the telco established organizationally exposed strategic areas for Energy, Automotive and Health, to create a supporting environment for present and following radical non-core-telco related innovation projects. The strategic areas organizationally were placed in the B2B entity structure of the telco group and not under the group's governance, because they predominantly developed solutions for business customers. It was aimed that the strategic areas were able to act as independent as possible from the telco groups processes, to act quickly and flexible, following a so called speed-boat approach. They got equipped with innovation budget for its own disposal, but also were dependent on group's resources and had to fulfill mid-term financial expectations, regarding generating significant revenues with new businesses.

When the strategic areas got established, the energy and automotive projects got assigned to the respective strategic areas that, from that time, steered and governed the further innovation activities. The energy project got hardly affected by the new organizational conditions, because the platform nearly reached its basic scope of functionality and necessary budgets already got released within the product development process. However, radical changes affected the automotive project, because the respective strategic area decided not to move forward in the given constellation. It only fulfilled promised agreements and focused on other innovation projects.

The health project was conducted entirely in the new strategic area environment and leaner processes applied. After an in-depth assessment of the project proposal, its budget was quickly and relatively non-bureaucratically released by the strategic area health and a core project team has been set in place. However, the team was dependent on various group resources; amongst others further experts, procurement, legal conformity declaration, SAP-adaptation. The dependency on other group entities applied to all project teams before and after the projects got assigned to the respective strategic areas. Hence, it was necessary to follow various group processes and to pass processes as well. To minimize its organizational workload necessary for passing gates and following processes, the strategic areas established strong ties and interaction channels across the whole telecommunication company group, but despite all efforts, the formal requirements to make use of the group's resources were considerably high.

5.4.3 Environmental Rigidity

From the first described innovation project that started in the end of 2007 to the creation of the case studies in the end of 2012, the telecommunication company's organizational structure and processes changed a lot, various gate processes were less present and the teams conducting innovation projects were able to act more independent and had more degrees of freedom. However, its dependency on group-wide resources hindered and slowed down its speed of action significantly and accordingly its speed to innovate.

It turned out that the aimed speed-boat approach with a focus on acting independently from the group's processes was not established to its full extent and it was highly complicate to realize it. One of the main reasons for the rigidity of the extensive processes and formality was that the group-wide process owners and managers who were responsible for the required resources did not accept that approach. On the one hand they had defined these long-standing processes and feared that leaner processes might reduce their internal power and reputation, hence, they had no advantage in allowing fast tracks or other special solutions for the strategic areas and/or its innovation projects. On the other hand, their personnel compensation and incentive systems were customer project oriented. Against that background it usually made more sense for them to support projects with short-term revenue expectations. Accordingly they had no interest in changing their processes, to release rare experts or to support the strategic areas in an alternative way.

The awareness grew in the strategic areas that if a project or idea was accepted by their internal management or even by the groups executive board, it did not implicate that the organization as a whole supported the decision, rather responsible managers insisted on the requirements to pass their boards and processes. Potential adaptations only were possible after intensive bilateral efforts between the respective process owner and the respective head of the strategic area, depending on the goodwill and benevolence of the process owner. Sometimes it was possible to establish fast-tracks, but usually the processes were very similar to the general processes that were relevant for much bigger entities and projects, what made it hardly possible to follow the speed-boat approach within the company's boundaries. Administrative group-internal efforts took a major part of the activities of strategic areas management level. Even new processes had to be established, e.g., when an innovation project reached the commercialization phase of the solution and many entities were involved to bring the new solution to the market. In this phase internal systems had to be adapted to integrate the solution in the portfolio and to provide all entities needed accesses to information. Moreover, the sales target figures based compensations and incentive systems of the sales force needed to be adapted to stimulate their sales efforts, what took time and persuasiveness, and delayed the effective go-to-market. Due to the fact that one of the most important success factors for innovations was the quickest possible time to market, the speed-boat approach was a first step in the right direction, however, with areas of improvement regarding integration in the companies' structure and processes.

5.4.4 General Conditions

Without a major supporter with reasonable financial strength and political power within the company, innovation projects were not possible, as observed in the automotive project, where the strategic area decided not to push the development. On the other hand, even the group's executive board support was not sufficient to guarantee the adoption within the company, as seen in the energy project. Hence, the innovation projects needed the support of all affected managers and process owners, across the whole group. Extensive processes and gate models were common and partly needed in big companies, across all industries, to guarantee regulated decision making, effective market approaches and to steer and support their core business activities, but cross-industry innovation projects did not fit this culture and needed leaner processes that were not established at the time of conducting the case studies. The involved small companies were much more flexible and faster in decision making, because they followed leaner or no defined processes, even though they did not follow defined innovation processes.

The telecommunication company had a large workforce with a lot of know-how and well-rehearsed teams with long lasting experience. The applicable specialists increased the chances of success of innovation projects, however, for the specifics of cross-industry innovation projects industry experts were quite rare. The strategic areas build up expertise related to their respective industries, but it took time and the required know-how was not applicable in the beginning of the projects. Hence, cross-industry partnering was the only possibility to develop the desired solutions. Partnering was common for the telecommunication company before the described projects, but despite its innovation entity's activities not in terms of developing new solutions, rather more in terms of hardware supply.

Depending on the earlier mentioned industry characteristics, the innovation approach and related innovation processes differed a lot across the companies.

They were either well defined or did not exist. Open innovation characteristics have been considered, in the innovation processes of the telecommunication and the automotive company, however, they differed regarding the fact that for the automotive company it was common to release a considerably amount of innovation budget and invest it to develop a topic even without a pilot customer, especially in contrast to the telco's B2B approach. The energy company predominantly reacted on changes in legal regulations and bought existing solutions. Cross-industry characteristics have not been consciously considered in any of the companies' innovation processes, because for all involved companies and its people, cross-industry innovation projects were a new approach to innovate. Some people partnered before with its suppliers or had projects with customers from another industry, but no one was involved in a cross-industry innovation project.

6 Analysis and Discussion of Findings with Existing Literature

Literature review is another source of data, generally conducted in the sorting phase of notes that supports the development of a theory. Following Glaser & Strauss (1967) it relates first findings and notes with existing literature (Dunne 2011; Glaser & Strauss 1967, p.37). Guided by the evolved concepts and overarching categories this chapter analyzes first findings and gives new impulses for the emerging theory based on literature review (Eisenhardt 1989). Both support to formulate more precisely practical strategic implications for telecommunication companies conducting cross-industry innovation projects.

6.1 Market and Industry

Each company is adapted to its industries needs and specifics. In cross-industry innovation projects it is essential to attune to common requirements and to understand the whole target industry context.

6.1.1 Innovation Speed and Approach

The cross-case findings in this study (cf. chapter 5) identify different innovation speeds across the industries. This is in line with Roberts (2007) who notes that the process for industrial product innovation generally takes three to eight years from initial idea to market, but it also can take up to 30 years (Roberts 2007). Based on industry-specific judgment, among partners from different industries, the same innovation can be considered as disruptive for one partner, while another might only find it incremental (Enkel & Gassmann 2010). This also reflects in the heterogeneity of innovation processes across industries (Leiponen & Drejer 2007). OECD and Eurostat (2005) note that innovation processes differ massively in terms of development, rate of technological change, linkages and access to knowledge, as well as in terms of organizational structures and institutional factors (OECD & Eurostat 2005). This suggests that innovation partners from different industries need some time in the beginning of the collaboration to adapt and align their working and innovation speed, and to understand the other innovation process, to start an effective and efficient collaboration.

In the high-technology industries, such as the electrical, electronic and IT, R&D plays a central role in innovation activities and the number of joint R&D projects comprises almost 50 percent of all R&D projects within a company. While slow clockspeed industries rely to a greater degree on the adoption of knowledge and technology and the number of joint projects is 20 percent or less (Enkel et al. 2009; OECD & Eurostat 2005). This might explain why it was challenging to identify innovation projects in the telecommunications industry with partners from a non-ICT industry.

The implementation of cross-industry innovation processes will lead to more radical innovation compared to innovation within one industry because the knowledge-base increases, allowing to identify analogies and the adaption of distant solutions. However, the present case studies reveal that efforts to realize radical innovations come along with challenges in present organizational structures, especially because present incentive systems are not in place to support radical innovations (Grote et al. 2012). Roberts (2007) and OECD & Eurostat (2005) note the relationship between innovation process and organizational structure. The authors identify that differences in innovation activity across industries, whether mainly incremental or radical innovations, also place different demands on the organizational structure of organizations, and institutional factors such as regulations and intellectual property rights can vary to a great extent in their role and importance (Roberts 2007; OECD & Eurostat 2005). This suggests that the present study also needs to respond to questions regarding organizational requirements after identifying a practical process model for cross-industry service innovation projects (cf. chapter 7).

6.1.2 Commercialization and Customer Acceptance

The value of innovations goes beyond the impact on the developing organizations. Companies consider external conditioning factors, especially the cost and risk of innovation related to market acceptance, as more significant barriers than internal barriers, such as the lack of qualified personnel and organizational rigidities (Howells & Tether 2004). Hence, it is necessary to examine the effects and benefits of innovations for other organizations, consumers and the general public (OECD & Eurostat 2005). The interviews reveal that market acceptance is one of the key challenges in the projects and confirm academic evaluation that new product development involves uncertainty about its market adaption (Jiménez-Zarco et al. 2011). The high failure rate of innovation projects is primarily caused because customers do not adopt the innovations. The reasons may be minor, but often are results of insufficient knowledge of customers' preferences and requirements (Ribiere & Tuggle 2010; Newman 2009; Martin & Horne 1995). Organizations should establish cooperative relationships to reduce uncertainties while creating offerings in line with market needs and to improve the overall project success (Liao 2001). Moreover, knowledge about customers environment enables the organizations to offer creative solutions (Jiménez-Zarco et al. 2011). In addition, the success of an innovation often depends on other external innovators in its environment, that supply additional complements or components (Adner & Kapoor 2010). Innovating organizations should intensively try to understand the whole target industry context to add desired components to its portfolio and to potentially provide a whole ecosystem of complementing services, most probably in consultation or collaboration with companies within the target industry. This increases value for the customers of the new solution and the aimed market adaption. Confirming these thoughts, the described projects illustrate that even following a cross-industry innovation approach does not guarantee market success, and especially customers' demands and thoughts need to be considered to a great extent. Gales & Mansour-Cole (1995) concern that large numbers of potential customers increase rather than decrease the uncertainty (Gales & Mansour-Cole 1995). This suggests that the innovation team should just try to win one or a few pilot customers and should not start other collaboration and commercialization activities until the innovation project reaches a certain maturity.

6.2 Innovation Project

This chapter relates first findings around the category 'project' with existing theoretical research and focusing the development of practical implication for telecommunication companies.

6.2.1   Pilot Customer and/or Innovation Partner

'Market pull' more frequently leads to successful innovations than 'technology push', although both sources of initiating projects account for success and failure alike (Newman 2009; Roberts 2007). However, improving market research and just listening to the customers is not sufficient for radical innovation development, rather a deeper analysis is necessary. "Henry Ford was fond of saying that if he had performed market research prior to developing the automobile, the responses he would have received would not have pointed toward the need to develop automobiles but rather toward the invention of faster horses that ate less hay." The question need to be answered: What is the job that the customer is really trying to get done? (Christensen & Euchner 2011; Ribiere & Tuggle 2010)

Carbonell et al. (2009) identify that across all stages of the innovation process, customer involvement has a positive direct effect on technical quality and innovation speed, and an indirect effect on competitive superiority and sales performance (Carbonell et al. 2009). Gales & Mansour-Cole (1995) add that customer involvement may help the organization to gain control of scarce resources, gain power, manipulate potential customers through cooptation, or fulfill role obligations, and hence, reduce uncertainty (Gales & Mansour-Cole 1995). Consequently, in accordance with the case study analysis (cf. chapter 4 and 5) which proposes that a pilot customer increases successful market positioning and commercialization activities, and Gemünden et al. (1992) findings that the relevance of information from customers is strongly correlated with technological innovation success (Gemünden et al. 1992), it suggests, that customers should extensively be involved as tight as possible in cross-industry innovation projects. The customer involvement should even cover the initial stages, like the idea generation and screening states, to critically evaluate the concept of the planned solution (Alam 2007; Martin & Horne 1995).

Though, the organizational condition of the customer interaction need to be considered in particular: focusing on seller-buyer relationships, Athaide & Zhang (2011) identify that the development of customized innovations calls for a product co-development relationship, while a discontinuous innovation, requires an emphasis on unilateral education based relationships (Athaide & Zhang 2011). Grant & Baden-Fuller (2003) emphasize the distinction between knowledge generation (exploration) and knowledge application (exploitation) goals of organizations, and they argue that alliances that follow a common knowledge accessing approach are more stable compared to alliances where knowledge acquisition is in focus (Grant & Baden-Fuller 2003). Athaide & Zhang (2011) findings may be transferred to cross-industry innovation projects where organizations are dependent on industry specific knowledge and may be a good reference to determine to what extent the pilot customer takes the role of an innovation partner at the same time, and in specific if an additional pilot customer project needs to be conducted in parallel to the innovation project or if it is conducted as the nucleus of the innovation. That additionally might be one explanation why the automotive project did not get the required support within the strategic area. It was stable because the innovation partners had a common knowledge accessing approach, and moreover, a pilot customer for unilateral education was not available.

6.2.2 Innovation Project Phases and Activities

The case studies illustrate that despite the commonly accepted sequential innovation process, the phases were partly overlapping and some activities spread across the whole project, suggesting the assumption that accepted innovation process models assuming predominantly tangible product development, find their limitations in service innovation (Ottenbacher & Harrington 2010). This observation clarifies Roberts (2007), who notes that process models may not show feedbacks from later stages back to earlier ones due to simplicity sake. He adds that these feedbacks among activities inevitably exist in real processes and cause reiteration among the stages, with major variations to specific tasks, managerial issues and managerial answers throughout the whole process (Roberts 2007). Weck (2006) and Alam & Perry (2002) give careful consideration to linear and parallel stages of an innovation process and develop a model with overlapping activities. They suggest that managers should establish a linear innovation process and conduct some stages in parallel to fast track the overall process, namely strategic planning and idea generation, idea screening and business analysis, personnel training and service testing (Weck 2006; Alam & Perry 2002). This approach is very similar to the observed stages and activities in the case studies, namely continuous ideation, market analysis and commercialization activities in parallel to the interlinked innovation project phases.

Froehle & Roth (2007) give a good overview of possible practices and activities that occur across the different phases in innovation projects. They define 45 practice constructs covering resource-oriented as well as process-oriented practice constructs and address topics such as formalization, allocation, communication and commercialization (Froehle & Roth 2007).

Ideation generally is done within the organization and after evaluation and assessment organizations might consider partnering with externals (Magnusson 2003). This is a logical approach, however, Russo-Spena & Mele (2012) suggest that activities in which actors interact, collaborate and integrate their resources, should be integrated in each phase of the innovation process, following a five Co-s model that comprises (not necessarily sequentially) co-ideation, co-evaluation, co-design, co-test and co-launch (Russo-Spena & Mele 2012). Moreover, formulating problems or new solutions from different perspectives and at different levels of abstraction enhances creativity. Inter-domain conceptual combination (far association) is more conducive to generating creative ideas than intra-domain conceptual combination (near association) (Zeng et al. 2011). This suggests that innovation teams should discuss the initial ideas with the pilot customer and/or innovation partner and make use of their high innovativeness at an early stage in the innovation process (Magnusson 2003). E.g., a demonstrator was part of all three projects, differed in its scale and was a good way to visualize a common understanding (Thomke 2001). In case a specific problem is identified, the TRIZ method describes the way via the abstraction of the problem and an abstract analogue solution, to a specific solution in another field; Mann (2001), Moehrle (2005) and – considering the specifics for service design – Su et al. (2008) and Chai et al. (2005), give a good overview of the inventive problem solving approach that uses knowledge from former inventions and potentially leads to interesting solutions in distant industries (Su et al. 2008; Moehrle 2005; Chai et al. 2005; Mann 2001).

Regarding fuzzy front end activities of innovation projects, the extent to which a formal search for innovative ideas can be implemented is limited. Formal structures, systems and strategy should rather provide an environment, that fosters the possibility for new ideas to emerge anywhere in the organization and also to gather momentum through informal search activities (McGuinness & Conway 1989). In the transition phase of fuzzy front end and development, the case studies notice an increased collaboration intensity with the partner organizations, to define the structure and scope of the collaboration and to agree on working packages. From that point the organizational context and in particular the inter-functional integration across all process stages of new service development play an important role (Perks & Riihela 2004), as well as common learning activities, that occur throughout the process at individual, group and organizational levels (Ritala et al. 2009; Stevens & Dimitriadis 2004). Following commercialization decisions can influence consumer acceptance (Chiesa & Frattini 2011). Hence, an aligned approach is necessary and all innovation partners should agree to the planned commercialization activities, preferably in the beginning of the collaboration. The importance of aligned activities across partner organizations throughout the innovation process makes it inevitable that the partner organizations fit in terms of strategic goals, working habits and approaches (Schreiner et al. 2009; Bach & Whitehill 2008). Consequently, organizations should make sure that the identification of 'the best' partner is guaranteed, what may be formalized in the innovation process model, as an additional phase of partner finding. Compared to open innovation models described in literature, an additional phase of partner finding, will take the specifics of cross-industry innovation service projects into account and facilitate the successful realization.

6.2.3 Project Management

Innovation is a multi-stage process, with major differences needed for effective management of each stage of activity (Roberts 2007). Managers should understand that each phase provides an opportunity for collaboration that can enhance the value of the co-creation process (Russo-Spena & Mele 2012). The three main tasks of open innovation professionals across all phases consist of managing the inter-organizational collaboration process, managing the overall innovation process and creating new knowledge collaboratively (Du Chatenier et al. 2010).

Proper definition of management of interfaces is important in maintaining good working relations, especially when partners are unfamiliar with each other's business logic, organizational structure and processes (Gassmann, Zeschky, et al. 2010). The case studies reveal that non-hierarchical decentralized modes of organization and regular open confident communication facilitate strong inter-organizational linkages (Teece 1989). The development of an interfacing role is essential to manage and communicate expectations and tasks with the external parties and to solve potential occurring problems (Perks & Riihela 2004). This role also is essential among internal parties, because an increased involvement of external parties complicate internal cross-functional collaboration, caused by a complex web of various stakeholders making its management more challenging (Mortara et al. 2009, pp.30–40; Perks & Riihela 2004). A practical solution was illustrated in the case studies in terms of a steering board for balancing internal expectations with the ones of the external partner, as well as an innovation board that balances general internal inter-functional challenges. These instances establish a clear and structured hierarchy, and help the project team to focus on the needs of the actual innovation project.

Eisenhardt & Tabrizi (1995) note that product development is a process of navigating through unclear and shifting markets and technologies. The authors identify a correlation between fast product development and the experiential strategy of multiple design iterations, extensive testing, frequently project milestones, powerful leaders and multifunctional teams. The authors suggest using experiential and improvisational tactics as well as a more real-time, hands-on approach to realize fast product development (Eisenhardt & Tabrizi 1995). This reflects statements in various interviews conducted for the case studies, assessing that processes should only be guidelines for actions in innovation projects. Rather the projects need to be detached from common extensive processes of telecommunication companies and need the degrees of freedom to act in start-up-like environment.

6.3 Cross-Industry Collaboration

People with its collaboration characteristics are the biggest challenge and the biggest potential at the same time, while conducting cross-industry service innovation projects.

6.3.1 Partner Selection

In radical service innovation projects, managers should focus on the early stages, in particular idea generation, idea screening, team formation and knowledge acquisition, because they leverage the success of the whole project (Schweitzer & Gabriel 2012; Alam 2007). Especially the team formation of cross-organizational teams does generally not experience the required attention, e.g., the energy company and its associated IT-companies got the telecommunication company's innovation partner by chance, and the trustworthy relationship between a couple of automotive company and telecommunication company people was the reason why the organizations started the project endeavor. The case studies reveal at various points that partner selection played a minor role even though cross-industry innovation projects are determined by the fit of the partners to a great extent. Especially in radical innovation projects, where the amount of specific knowledge is high, collaboration quantity plays an important role (Schweitzer & Gabriel 2012), and partner match has a positive influence on collaboration competency (Tsou 2012a). The right people need to be integrated at the right time, using the right integration methods, and the right spirit of collaboration should prevail (Schweitzer & Gabriel 2012). Consequently, an in-depth analysis, both from a technological and contextual perspective is important to evaluate what knowledge is necessary to transfer (Gassmann & Zeschky 2008). A subsequent abstraction of the desired solution facilitates discussions with cross-industry partners. As a result and in contrast to the case study approaches, beside the need to choose a partner because of complementary resources, technologies and markets, also the need to choose a partner because of sharing the same values is an important premise (Bach & Whitehill 2008; Miles et al. 2006).

Cognitive distance among partners is predominantly regarded as a threat instead of an opportunity and advantages due to heterogeneity among innovation partners have not been researched extensively (Enkel & Gassmann 2010). While the need for cognitive proximity is important in traditional innovation projects that focus on incremental innovation within one industry, such as seller-buyer product co-development, where prior relationship history has a positive and statistically significant impact (Athaide & Zhang 2011), cross-industry innovation projects benefit from the partner specific knowledge and approaches to a much greater extent. In fact innovation performance is a parabolic, inverted u-shaped function of cognitive distance, in terms of differences in technological knowledge, between the partners and the effect is positively related to the innovativeness of the collaboration (Nooteboom et al. 2007). Accordingly, a minimum level of 'fit' is essential, related to common objectives and expectations, organizational structures and cultural aspects, however, too strong linkages may result in collective inertia with less innovativeness (Gassmann, Zeschky, et al. 2010; Eisingerich & Bell 2008).

A common goal can outweigh shortcomings in other areas (Gassmann, Zeschky, et al. 2010), but due to the fact that cross-industry innovation collaboration is generally accompanied with high cognitive distance among the partners, it consequently, is important to deeply analyze potential partners (Tsou 2012a). Among others, the partner analysis process also should cover a detailed analysis of the potential partner's business model, because entering new markets often require new business models that supplement or complement the original business models of all involved partners. Consequently, it is necessary to identify each company's business driver and areas of uncertainty what allows each side to assess the value of the relationship (Hummel et al. 2010). Moreover, it is necessary to aligning and formalize the expectations of the partners on intellectual property in this early phase of the collaboration and managers should respect a fair distribution of all benefits and that partners complement each other (Bstieler 2006). In this way, uncertainty can be converted into manageable risk. Moreover, a common understanding of each other's business models and a common go-to-market model, allows decreasing the overall risk by allocating respective tasks to the partner that can deal with it in the most appropriate way (Hummel et al. 2010).

Additionally, during the partner selection process, organizations should particularly focus on market leaders, because they provide key complementary resources and a collaboration signals importance and future prospect of the alliance. Hence, firms should try to attract market leaders in order to take advantage of the wealth spillover (Han et al. 2012). To sum up, managers need to be aware of the major influence of a well-match. They need to clearly define the project objectives and explore the mutually of their interests, to identify appropriate partners (Tsou & Chen 2012). Selecting partners carefully reduces the risk of failure (Gillier et al. 2010), because collaboration competency and partner match relate positively to knowledge integration mechanisms, which in turn relates positively to service innovation (Tsou 2012a). This especially applies to cross-industry innovation projects where collaboration is even more important than in traditional innovation projects. Hence, it seems advisable to explicitly define and stress the phase of partner selection in theoretical and practical innovation process models.

6.3.2 Contractual Structures and Trust

The case studies reveal that contractual frameworks are firmly established in organizational environments, primarily to reduce uncertainty that inevitably exists in collaboration with structurally different partners (Gassmann, Zeschky, et al. 2010). Contracts are an approved way to align both parties' interests and goals in a mutually acceptable way, to regulate intellectual property rights and to foster a mutual trustful understanding (Gassmann, Zeschky, et al. 2010). However, the strength of the partnership is not defined by contracts, it rather is a (probably linear) combination of the amount of time, the emotional intensity, the intimacy (mutual confiding) and the reciprocal services which characterize the partnership (Granovetter 1973). Contracts often remain incomplete, because actions and results cannot always be anticipated, especially in the cross-industry innovation context, and partner preferences also can evolve during the project, implicating revisions of the initial conditions, to maintain common interests (Gillier et al. 2010). Formulating contractual documents more detailed, accordingly does more hinder than support the collaboration. Partners rather should focus on a general agreement how much resources and what kind of knowledge each partner ensures to bring in and how each party participates in the commercialization of the solution. The smaller the amount of committed capital is, the looser the governance structures are required (Teece 1989).

Companies that acquire knowledge from partners through private channels, no matter if they are formal (official collaboration projects) or informal (chat at a conference), take less action to regulate knowledge outflow in contrast to acquired knowledge via public channels (patents, publications, press releases), they rather facilitate inter-firm knowledge exchange (Häussler 2010). This fact was also confirmed by most interviewees: after a phase of alignment, relations on personal level got tighter and a common sense and trust evolved, that facilitated knowledge exchange. Consequently, collaborating organizations should make use of multiple procedural, human, and organizational 'bridges', to link information carriers, via information collectors, to information translators and supporting actors/facilitators (Gottfridsson 2012). The roles and functions should collaborate intensively via private channels (face-to-face meetings, telephone calls, e-mails). Human bridges are the most effective transfer mechanism, and people movements, rotations and face-to-face meetings should be used routinely and frequently (Roberts 2007). This approach fosters communication, information sharing and the creation of trust.

Beside strategic, organizational and operational aspects, trust is an important factor to be considered, especially because uncertainty cannot be eliminated completely (Getha-Taylor 2012; Gassmann, Zeschky, et al. 2010). Trust and interaction on the individual level creates a basis for know-how transfer and reduces the risk of opportunistic behavior (Kale et al. 2000). Trust and know-how transfer are mutually dependent: communication through information sharing and perceived fairness, both procedural and distributive, are positively linked to trust, while continuous conflicts and egoism reduce trust (Bstieler 2006). Trust also allows to exposes the partners to each other's thoughts, improves the harmony within the working relationship and supports the planning and collaboration process. It is positively linked to project performance and a central element for collaborative development. Accordingly managers should emphasize a timely, reliable and adequate information exchange what is essential to realize a trustworthy relationship (Bstieler 2006). It needs to be considered, that trust building is a time-consuming process, and that the development of collaborative capabilities requires not only learning new values and behaviors but also unlearning old habits (Miles et al. 2006). An open mindset of all participants is the basis for a culture that values outside competence and know-how. This culture is influenced by the values of the company and many concrete artifacts such as incentive systems, management information systems, communication platforms, project decision criteria, supplier evaluation lists, etc. (Grote et al. 2012; Gassmann, Enkel, et al. 2010; Ritala et al. 2009).

6.3.3 Methods and Tools to Improve Collaboration

Working in an open innovation context typically may include challenges of high diversity, low reciprocal commitment, power differences and cognitive distances, high level of uncertainty including unsafe learning climate, low resource availability, absence of traditional hierarchical lines and insufficient absorptive capacity (Inauen & Schenker-Wicki 2011; Du Chatenier et al. 2010). To counteract these challenges, managers should identify areas in which their organization may be lacking inter-firm development competency and develop specific capabilities, foster an alignment of knowledge and technology integration mechanisms and encourage frequent interaction and close collaboration to maintain stable and enduring relationships (Tsou & Chen 2012; Getha-Taylor 2012; Newman 2009; Ritala et al. 2009). They should focus on individual competences and select and integrate the right people at the right time to save time and energy and to reduce costs (Ritala et al. 2009; Stevens & Dimitriadis 2004). This may be in coordination with HR staff based on their competence profiles (Du Chatenier et al. 2010). Due to the fact that brokering and social competence is highly important in the complex social and communicative environments of open innovation projects, training should be a continuous activity due to often scarceness of time. Moreover, managers should interact with the sales team in an early phase of the project to facilitate a better understanding of market, customers and business environment (Frisanco et al. 2008). If conflicts arise managers should not follow an egoistic behavior and stick to contracts, rather they should find a solution that creates a win-win situation for all parties and inspires the relationship (Bstieler 2006; Weck 2006). Even though managers follow traditional management methods in general, their obedience with intuition is more important in cross-industry innovation projects, because of the characteristics of the respective partners. Du Chatenier et al. (2010) identify brokering solutions and social competence as the most important competencies to deal with open innovation project challenges (Du Chatenier et al. 2010). Project leaders with knowledge broker skills that foster developing a common background while using information systems, enhances the partners communication capabilities (Gillier et al. 2010).

Technologies help to establish and improve communication and collaboration processes even when the team members' physical and cultural distance might be high. They accelerate the knowledge diffusion process, support to build a consensual vision (Gillier et al. 2010), reduce the perceived distances between the actors of the innovation process (Gassmann 2006), and are potentially more cost-effective (cf. video-/telephone-conferencing) compared to physical meetings (Telstra Corporation Ltd. & KPMG International 2012). There is a positive effect of information and communication technology on collaboration in new product development projects (Jiménez-Zarco et al. 2011), and proper systems for knowledge management and integration can improve competitiveness and innovative ability (OECD & Eurostat 2005). The systems take an enabling role in innovation projects and its integration, preferably including a cross-organizational alignment of processes, enhance the managers co-development competences (Tsou & Chen 2012). However, Ordanini & Parasuraman (2011) cite Marinova (2004) that knowledge acquired from outside the organization, often does not become available for innovation purposes due to inadequate mechanisms for integrating and sharing the information throughout the organization (Ordanini & Parasuraman 2011; Marinova 2004). Consequently, cross-industry innovation project teams need access to sophisticated knowledge integration mechanisms from the start, consisting of formal processes and structures that facilitate the capture, analysis, and synthesis of various types of knowledge and the distribution of that knowledge within the team and across the organization. Likewise, state-of-the-art collaboration tools are of outstanding importance in cross-industry innovation projects, because its perceived usefulness and compatibility have a significant impact on the project outcomes (Plewa et al. 2012; Grote et al. 2012).

6.4 Organizational Environment

Each participating organization needs to establish an innovation friendly environment that fundamentally differs from supporting the development of a groups existing product portfolio.

6.4.1 Processes and Structure

The organizational structure of an organization affects the efficiency of its innovation activities. A greater degree of organizational integration improves the coordination, planning and implementation of innovation strategies. Accordingly, organizational integration is recommended in industries characterized by incremental changes in knowledge and technologies, while a looser, more flexible form of organization, with more autonomy for the team members in decision making and defining their responsibilities, is more effective in more radical innovation environments (OECD & Eurostat 2005). Likewise, the structure of the innovation project and the innovation itself are interdependent. The innovation project structure is content-dependent, needs to evolve and adapt during the innovation process and determines coordination costs. The phases of the innovation process affect the coordination requirement (Linnarson 2005) and include a number of in-house activities that are not included in R&D, such as the later phases of development activities and implementation activities for the adoption (Roberts 2007; OECD & Eurostat 2005). Accordingly, the alignment of internal and external structures and processes, both inter- and intra-organizational, is a key factors to improve efficiency and success rate in cross-industry innovation projects and facilitates the successful integration of open innovation activities in internal innovation processes (Brunswicker & Hutschek 2010; Chien & Chen 2010).

An organizations ability to collaborate with other organizations starts from being able to collaborate internally (Miles et al. 2006), because innovation require complex, vertical, horizontal and lateral interactions and even integration among various organizational units (Mortara et al. 2009, pp.30–40). These interactions demand special infrastructure and processes to proceed smoothly and efficiently and no hierarchical internal structures that suffocate the innovation process (Teece 1989). The main challenge for cross-industry innovation projects typically are mid-to long-term commercialization goals for radical innovation and at the same time the dependence on corporate resources. The character of these projects inevitably leads to internal resistance, because it may cannibalize the present product portfolio, opposes present incentive systems, and they need start-up-like processes for budget and resource allocation, that contradict with corporate structures and processes (Enkel & Goel 2012). Even central R&D units may be reluctant to support, because they typically focus on short-term, customer-oriented business unit research activities (Gassmann, Enkel, et al. 2010). Companies typically fear too much openness, because it could lead to a loss of control and core competences, with a negative impact on the companies' long-term innovation success, and hence, they invest simultaneously in closed as well as in open innovation activities and retain extensive structures and processes (Enkel et al. 2009).

Innovation projects complexity increases with its innovativeness and time to market is a key competitive element, hence, cooperative relationships are needed to receive necessary assistance and knowledge to launch the new product on time (Jiménez-Zarco et al. 2011). A major mistake is to set up stringent formal processes in the early stage of the innovation process, for approval of relatively small amounts of R&D budget needed to try out an idea and for resource allocation. These heavy and discouraging evaluative procedures should rather be replaced by first- or second-level R&D supervisors with sufficient innovation budget to dispense at their own discretion (Roberts 2007). The legitimization and institutionalization of an autonomous division to perform radical innovation projects in the organization can help to protect the nascent technology from political opposition and to counteract other forces of inertia within the organization. It also increases the probability that the organization will commit early on commercializing the radical solution. In addition, an autonomous division can promote potential product cannibalization and help the organization to discount feedback from its existing value network (Hill & Rothaermel 2003).

6.4.2 Incentive and Support Systems

Collaboration is a complex, voluntary, self-managed and potentially fragile process, that can only be encouraged and facilitated, but not easily accelerated. The inherent rewards of collaborative interaction in a trusting, supportive community, that share identical goals and a good working atmosphere, are the driving force behind successful innovation projects. To be effective in the long run, teams must evolve at a pace that is comfortable for all of their members and in a way that members fully internalize and align necessary behaviors (Chien & Chen 2010; Miles et al. 2006). Personal objectives, incentives and compensation structure influence the behavior of transferring information and acting in or for the project regardless of whether it is in accordance with the economic interest of the organization or not. Hence, incentive schemes play an important role and need to be aligned to motivate employees to act in accordance with the project interest (Grote et al. 2012; Roberts 2007; Conway 1995). The case studies prove Christensen & Euchner (2011) finding, that business units that deal with traditional and new business respectively may face contrary incentive systems, that do not support the radical innovation project, and hence, need to be aligned or separated (Christensen & Euchner 2011).

Traditionally organizations implemented strategies that developed defensible positions against the forces of competition and power in the value chain, constructing barriers to competition, rather than promoting openness

(Chesbrough & Appleyard 2007). Top-management plays an important role, because their commitment to the open innovation approach is essential to assure success of broad-based programs aimed at institutionalizing the development of effective product and process innovations (Roberts 2007). They may initiate strategic developments in favor for cross-industry innovation, that implicate being open-minded for external solutions and the willingness to challenge or even cannibalize existing own technologies and portfolio elements (Gassmann & Zeschky 2008). They may establish open strategies and open business models based on innovation and coordination, that balance traditional business strategies with open innovation approaches. The new strategy needs to balance value capture and value creation, taking strategic sense of innovation communities, ecosystems, networks and their implications for competitive advantage into account (Chesbrough & Appleyard 2007). Huston & Sakkab (2006) believe that 'connect' and 'develop' will become the dominant innovation model in the twenty-first century, in contrast to traditional R&D approaches. The authors note that top leaders need to drive the new innovation approach and that "the CEO of any organization must make it an explicit company strategy and priority to capture a certain amount of innovation externally." Isolated experimental approaches in some organizational units or a solely strategic R&D responsibility will inevitably lead to a fail (Huston & Sakkab 2006).

7 Practical Strategic Implications

This chapter describes the strategic implications for telecommunication companies. Grounded and as an immediate consequence of the constituted categories project, company, collaboration and market, first findings got reviewed and developed in literature (cf. chapter 6), and thus, in this chapter illustrates the iteratively emerged theory.

7.1 Conduction of Cross-Industry Service Innovation Projects

The category 'project' emerged as core category during the analysis process and the other categories got aligned to it. Hence, guided by the research question, this chapter presents the process framework as a major part of the developed model. Organizational, cultural and portfolio topics extend and support the following practical process model.

7.1.1 Practical Process Model for Telecommunication Companies

Too structured and too formalized innovation processes may inhibit exploration efforts, that rely heavily on interactions among the partners. Hence, its necessary to balance the need for process and project structure with explorative encouragements (Kindström & Kowalkowski 2009). The following process model guides telecommunication companies while conducting service innovation projects with partners from non-ICT industries (cf. figure 15). In the following paragraphs the practical process model is described in detail. It starts with general internal steps and later considers two major variations related to the fact if a pilot customer is available or not. The former path variation considers if the customer project is part of the innovation project, and the later path variation considers the solution development solely with an innovation partner.

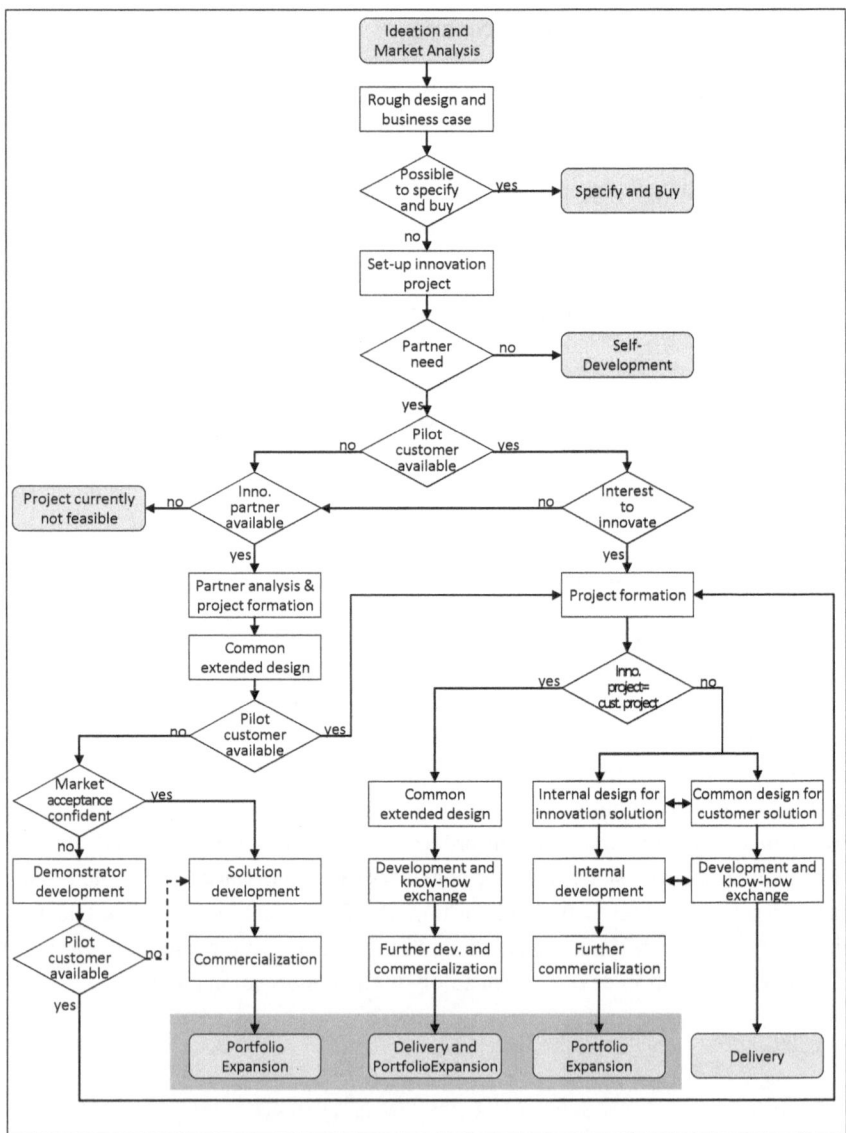

Figure 15: Practical Innovation Process Model for Telcos

7.1.1.1 Ideation, Analysis and Design

Conducting a new innovation project usually starts with **ideation and market analysis** (cf. figure 15). A new service idea typically either arises from internal continuous market research and analysis, or it is caused by a specific external event, such as discussions with representatives of other organizations. Also, internal idea competitions, technological announcements or selective trend analyzing literature are common. Every source of the new idea is appreciated, whether from external impulses, from within the organization, planed as daily business from strategy, business development or innovation departments, by employees from organized events or just by chance. In any case ideation and market analysis that includes cross-industry innovation component is both a process and a mindset, and the new service idea need to be analyzed in detail and reflected to technological feasibility, economical profitability and market needs (Nolf et al. 2012; Liao 2001). Therefore, a first **rough design and business case** is essential. Given the nature of cross-industry innovation projects a sustainable judgment usually is challenging but essential to decide if it is worth to proceed with the idea (Kleinknecht & Van Der Panne 2012). Sole unfounded support from high-ranked organizational members is not sufficient (Gemünden et al. 2007). In this step the initial team, probably in collaboration with internal or external experts, also identifies if it is possible to specify the solution in detail, and accordingly a make-or-buy decision is necessary, taking among others economic valuation and risk into account. In case it is not possible to specify the newly planned solution, the next step is to **set-up an innovation project**, found it with a sufficient amount of pre-budget to assess the potential solution in detail and with a core team fulfilling required roles, driving the new innovation topic (Mansfeld et al. 2010). The team needs to elaborate the existing technical and financial drafts, assessments and designs of the planned solution, and develop them further, and evaluate if an external innovation partner is needed, new experts should be hired or if the project can predominantly be self-developed as a traditional innovation project with the given workforce within the company's boundaries. If the company needs an innovation partner with non-ICT expertise, the team's first attempt should be to win a pilot customer that has an interest to innovate and is willing to support the development of the solution (Maklan et al. 2008). This assures that all property rights stay with the company, the team will not face the not-invented-here-syndrome and the developed solution is already positioned in the market for further commercialization activities. If no such pilot customer is available, the team needs to screen the market for potential innovation partner that are willing to contribute the required industry specific knowledge. Most important is a strategically fit between the partners, related to the exploitation of the developed solution (Gassmann, Zeschky, et al. 2010). If the team is not able to identify a suitable partner, intermediaries such as consulting companies with their networks might support, otherwise, the project is not feasible at that time and the company should consider building up internal expertise and/or realize it at a later time (Batterink et al. 2010; Howells 2006).

7.1.1.2 Project Formation and Conduction with Pilot Customer

After the innovation team has approached potential pilot customers and found one with an interest to support the innovation project, the parties need to agree on the practical conduction and the desired outcome of the project, what is part of the **project formation** step (cf. figure 15). It includes contracts with the pilot customer that script all questions among others related to project input resources, SLA and further commercialization. Within the discussions it gets obvious to which extend the customer project should be interweaved with the innovation project. It may be any state between being part of the innovation project, in terms of that it forms its nucleus, or that the customer project predominantly facilitates a smooth industry know-how transfer towards the internally conducted innovation project (Athaide & Zhang 2011). It may also define to what extend the pilot customer is willing to take the role of an innovation partner or of a traditional customer. In any case the innovation team needs to guarantee the internal resource allocation, received a substantial mid-term budget release for the solution development, set-up an internal innovation project structure and define clear responsibilities towards the customer.

The transition between an integrated and a detached customer project is fluent. In case most components that are developed in the customer project can be reused within the innovation project, and the customer solution forms the nucleus of the innovation project, the partners agree on a **common extended design** in the beginning of the collaboration. Beside the basic functionality of the customer solution it also includes aspects of the extended generic innovation approach. The integrated approach is preferred, because it is the cheapest and fastest way to conduct the innovation project. The pilot customer does not get deeply involved in the planning of the extensions, but need to get informed why additional time is required at some stages and why potentially some components are designed to follow standards instead of serving the customer specific situation. Within the **development and know-how exchange** phase, the team develops and tests the solution for the customer as earlier agreed. The phase is characterized by extensive know-how transfer from the pilot customer (and/or the partner company respectively) to support the development of the solution (Martin & Horne 1995). Without the input of the pilot customer (and/or the partner company respectively), the development of the solution usually would not be possible in an adequate time and with adequate budget (Carbonell et al. 2009). The non-ICT company also has an interest in the completion of the solution, and consequently, is willing to invest time and resources to provide industry specific know-how and support the development team wherever necessary. The development and test usually takes more time compared with the sole development and test of a customer specific solution, what takes the step **further development and commercialization** into account. Moreover, the development and test of additional modules and functionality might be necessary or appropriate to offer a sustainable solution in the market. Usually commercialization activities start while the development is still underway and the team extends the solution delivered to the pilot customer and anchors the solution as a new portfolio element within the company's structure.

In case the pilot customer insists on a tight schedule and/or various specific elements, the customer project need to be conducted detached from the innovation project and both projects run in parallel. One team faces to the customer and conducts the customer project, while another one conducts the innovation project. Every activity that is not part of the customer project is unknowable for the pilot customer and acts internal to develop the solution. Nevertheless, the teams are deeply linked and predominantly industry specific knowledge gathered by the customer solution team gets transferred to the innovation solution team, while, e.g., specific developed components get transferred the reverse way. As a first step the respective teams internally agree on an **internal design for the innovation solution** and in coordination with the customer on a **common design for the customer solution**, while discussions with the customer, inevitably influence the innovation solution design, because new industry specific information can be considered. In the next step, the innovation project team conducts the **internal development** and testing while the customer project team coordinates **development and know-how exchange** with the pilot customer. Again, industry specific expertise from the pilot customer that is needed to develop the customer solution inevitably influences and supports the development of the innovation solution. Whether two development teams for each solution exist, or the innovation project team develops the customer solution as well, the customer only is involved in the customer solution specifics, but the teams exchange newly gathered information, expertise and components. When the customer solution is delivered, the order is filled and **further commercialization** activities with other customers either start during or after the development. For the innovation solution the team might need to develop additional components for better market positioning. The team also approaches internal senior-, line-, and project portfolio managers, to push commercialization and to anchor the solution within the company's structure as a new portfolio element and need to consider additional topics such as privacy, legal assessment and product description to transfer the solution in a product status (Beringer et al. 2012).

7.1.1.3 Partner Finding and Project Conduction without Pilot Customer

If no pilot customer with an interest to innovate can be identified, the initial team needs to screen the market for a suitable innovation partner that is equipped with the needed non-ICT industry expertise. In the **partner analysis and project formation** process the team has in-depth conversations with potential partners and sounds feasible collaborations (Kale & Singh 2009)(cf. figure 15). In this phase it is indispensable, that it gets obvious what every partner brings in and desires to get out of the collaboration and if the partners are equal or not. The team defines criteria for the partner selection, e.g., related the size to guarantee reliability and need to keep an eye on a multi sourcing strategy. The teams should not just focus on technological and political facts, but more on a general personal and strategic fit between the organizations (Gassmann, Zeschky, et al. 2010; Enkel & Gassmann 2010). Even though the commercialization of the new solution is after the development, it still is part of the collaboration, and hence, the role of all partners in the value chain should be very clear from the beginning, otherwise, the risk of abortion gets more probable resulting in a loss of the earlier investments (Gassmann, Zeschky, et al. 2010). If the win-win situation is not clear, problems will arise while the conduction of the project, and hence, the process of mutual understanding should be conducted excessively. LOI and NDA need to be signed, because the projects affect strategically decision, and a partnership contract is important to verbalize the ambition of each of the partners (Gassmann, Zeschky, et al. 2010; Weck 2006). In general, the intention and motivation of the people is more important than contracts covering all contingency in it's entirely, but the basic aims need to be codified to guarantee the win-win situation (Granovetter 1973). A **common extended design** of a business model is part of the discussions and a common detailed design of the solution helps to get a better understanding of the common solution, to plan responsibilities, time frame and to set-up the project management. When the scope of the project gets clearer the initial team also should have reached a substantial mid-term budget release for the complete solution development. It might be possible that in the earlier process the initial innovation team identified a potential pilot customer that did not have an interest to support the development of the innovation, or the new innovation partner might have good relationships to potential customers. Hence, it should be checked again if a pilot customer is available, to improve the later market acceptance of the solution in the commercialization phase. In contrast to the earlier described 'Project formation and conduction with pilot customer' (cf. chapter 7.1.1.2) at this stage the potential pilot customer do not necessarily need to support the development, because the innovation partner is already equipped with the needed non-ICT expertise. If a pilot customer is available the process can be followed as described in chapter 7.1.1.2, otherwise, the team should judge the market acceptance of the planned solution. If they are confident that the market will accept the solution and a quick time to market is advisable, the team can start the **solution development**, as earlier planned in the design phase and later on the **commercialization** and integration in the company's portfolio and structure. This step requires higher pre-invest compared to the step with a pilot customer but might be wise in case time to market is important. High attention should be turned on the inter-personal relationships, tolerance and motivation of the cross-industry teams, what is inevitable for a successful collaboration (Tsou 2012a). If the team is not confident enough that the market will accept the solution, they should **develop a demonstrator** as a first instance that serves as a vehicle to visualize the planned solution and to convince potential customers to agree as pilot customer (Thomke 2001). In case the demonstrator successfully supported to win a pilot customer, the team proceeds the project formation and solution development as described in chapter 7.1.1.2, otherwise, it should be very carefully analyzed if it is worth to develop the solution without a pilot customer, considering among others market acceptance, investment volume, importance of time to market and market size.

7.1.2 Parallel and On-going Activities

While passing through the steps of the described process model (cf. chapter 7.1.1; cf. figure 15), various activities need to be conducted simultaneously (Weck 2006). Market analysis is a continuous activity not just throughout the project to assure aiming in the right direction, but for the company, identifying market opportunities, what is deeply interweave with ideation, the creative process of finding new ideas. Whether the new idea arises from an identified market need or technological possibility, it needs to be analyzed if technical feasibility or market acceptance respectively is given, and whether it is a strategic relevant and promising field for the company. Hence, a deeper technical understanding of the needed components is inevitable for a first business case, resulting from a first design that allows to roughly assess the development costs.

The internal business case constantly needs to be adapted, given new information and facts, especially while forming the project with an innovation partner, what generally includes a modified common design of the planned solution as well as a common business model (Koppinen et al. 2010; Bhattacharya & Krishnan 1998). The reassessment of the internal business case influences the innovation partner selection, because the omission of parts of the value chain or new industry specific information might decrease the attractiveness of the whole project. The internal business case also gets influenced by the role of the pilot customer. If the pilot customer is acting more like an innovation partner and allows developing major parts of its solution within the innovation project, the costs are considerably lower than compared to the situation where the pilot customer takes more the role of a traditional customer that expects its solution within a tight time frame, resulting in a parallel development of the innovation solution and the customer solution.

Establishing good relationships to various stakeholders, e.g., while getting part of industry-specific networks, associations and unions takes long time and big efforts. These groups are important while approaching a new market in a sustainable way (Hansen & Bunn 2009). In general, intermediaries have a strong industry-specific network and are beyond that a good starting-point to identify and approach potential innovation partners, pilot customers, future paying customers, and to build up alliances for a go-to-market. E.g., major management consulting companies have extensive networks across all industries and influence companies a lot in their decisions (Batterink et al. 2010).

Service development differs from tangible product development in terms that testing activities can be more easily integrated in the process, supported by sophisticated development environments, especially while following agile approaches that are appropriate for adaptations that usually appear in innovation projects within the development process (Segelod & Jordan 2004). In parallel to the development the team prepares a smooth transition into the company's structures, to anchor the solution as a new portfolio element and conducts commercialization activities. There is no rule when these activities start and usually are done in collaboration with respective experts or managers, and dependent on the contracts with the innovation partners.

Throughout the project a project controlling need to be implemented, including milestone reporting, to guarantee that the project is still on track regarding aimed outcome and financial indicators. In return the team needs absolute management support to reach conclusions with company-internal stakeholders more easily and to keep the people motivated (Mortara et al. 2009, pp.30–40). Especially in cross-industry collaboration projects the motivation to collaborate is inevitable for the success of the project. Hence, beside the internal support, the people should meet their external innovation partners as often as possible and collaborate as close as possible, to build up trust, tolerance, mutual understanding and commitment (Hansen & Bunn 2009; Hoegl et al. 2004; Hoegl & Gemuenden 2001). It might also be useful to revert to intermediaries such as external consultants or experts in any state of the process, for market or technology analysis, project formation, or to use their relationships to approach the market (Batterink et al. 2010; Howells 2006).

7.1.3 Overarching Project Structure and Management

Beside a project manager, the innovation project has a project team that is growing in size with the maturity of the project and gets support, whenever needed, by internal and external experts. Sub-project manager with professional expertise are also advisable to steer the specific tasks. Whether the same project manager or team conducts a potential pilot customer project as well or if they appoint other people, is variable. In case they collaborate with a partner company they distinguish very clearly between the internal innovation project and the external collaborative innovation project, and focus on the internal business case, confidentiality matters and the company's strategy, while driving the common project at the same time. A steering committee consisting out of the project managers and top-managers, preferably from an internal innovation board, representing all parties is an effective means to solve potential disagreements.

First ideation and market analysis approaches are founded by the respective entity that initiated the advance. When the first analysis results in the set-up of an innovation project due to an innovation board decision, it gets staffed with a core team and equipped with sufficient pre-budget of dedicated innovation budget to analyze and assess the opportunity in detail and develop the concepts further. A substantial mid-term budget release for the solution development is then released by the innovation board in the project formation phase, that also includes the formation with the partner company and/or pilot customer. The budget release need to be sufficient to develop the solution and provided in tranches whenever needed. Reporting to the innovation board is necessary to guarantee that the project is still on track, but not to apply for the tranches and should be reduced to a minimum.

Project managers should pay particular attention both to the interorganizational relationship and to project management (Couchman & Fulop 2009). They are well advised to implement and follow traditional project management principles and methodologies, accompanied by supportive collaboration tools among others for content management, development environment, e-mail, telephone conferencing, and most important regular personal meetings (Grote et al. 2012; Roberts 2007; Weck 2006). Development-related planning and process measures, such as goal stability throughout the development process and proficiency of project risk planning, enhance the performance of innovation development significantly (Couchman & Fulop 2009; Salomo et al. 2007). The collaboration tools also support steering geographically wide spread sub-teams that carry out major working packages and are getting more and more common.

A challenge for the innovation teams is judging the importance of fulfilling demanded customer characteristics. In the energy solution project the team decided to use standardized interfaces that allow third parties to connect what made it necessary for the energy company to adapt its systems to connect to the solution of the telecommunication company. This approach was necessary to establish a marketable solution; otherwise, the service would have been too much focused on the specific customer's needs. Beside standardized interfaces, customers generally do not care about a scaling, multi-tenant solution that is one of the key requirements the innovation team demands. Hence, the pilot customer only can define parts of the innovation solution and the innovation team need to decide from the viewpoint of all potential customers, the needed set of functionality.

7.2 Organizational Preconditions and Requirements

While the process model predominantly emerged from the core category 'project', various organizational preconditions, concealed in the 'company' category, have been identified. This chapter lists the major principals to anchor the practical innovation process model within the company's boundaries, to guarantee fast decision making and to keep innovativeness on a high level.

7.2.1 Located within the Company's Structure

Cross-industry innovation teams of telecommunication companies need to combine the advantages of strong corporates and innovative start-ups while at the same time avoiding their respective disadvantages. Metaphorically, corporates are like oil tanker and start-ups are like speed boats. They differ among others in:

- innovativeness and processes – low, strict guidelines vs. high, just doing
- budget, business culture and mindset – rely on established cash cows, employees salary is safe, bureaucrats vs. innovate new products/services or the company gets insolvent, entrepreneurs
- compensation time frame and incentives – short-term revenue orientation aligned to present portfolio vs. mid-term product/service success orientation
- resources and reputation – all-round workforce, stable image vs. few experts, customers/partners are cautious.

Establishing a speed boat environment as a breeding ground for innovation, and hence, to increase success in cross-industry innovation projects, is either possible to realize in a legal external organization or in a self-contained project or entity embedded in the corporation's boundaries. It is not possible to just conduct the project in a traditional start-up structure, given the fact that the core innovation team is dependent on internal telecommunication experts and other resources, the brand image of the corporation to attract strong partners and/or pilot customers with good reputation, and moreover, the team needs to pave the way for a smooth transition of a high scaling solution in the corporations portfolio as soon as it reaches maturity. Likewise, it is not possible to conduct cross-industry innovation projects embedded in the existing structure, because all incentive systems are aligned to the core business, approval and decision processes are extensive, inflexible and time consuming, employees live the corporate culture and innovation is not part of their daily business (Bond & Houston 2003). Especially when the innovation team is part of the telecommunication companies B2B subsidiary/division that predominantly deals with customer projects, it is hard to establish an acceptance to spend big amounts of innovation money to develop a solution, particularly without a customer, and to allow fast decision making.

Hence, cross-industry innovation projects need to be conducted within the company's structure, but clearly separated from the operational business and as far as possible detached from its resource allocation and incremental innovation project processes (Enkel & Goel 2012; Buganza et al. 2011). The only authority the project manager needs to account for is an innovation board that is either located at board level or within the company's R&D/innovation department. The innovation board consists of top-managers whose personal compensations are not interwoven with the core portfolio elements of the company, and therefore, have an interest in innovating and not in keeping grown structures. It controls over a detached dedicated innovation budget and decides the set-up of innovation projects including a pre-budget and a potentially following substantial mid-term budget release for the solution development. The innovation board also needs to establish various ways to bypass or fast track given internal boards or processes that handle the allotment of company's resources, whether the innovation team is dependent on them and other shared-service infrastructure. This allows the core team to act relatively independent within the companies' structure, keeping reporting necessity and process formalities to a minimum, and still make use of the companies' amenities. In addition, no P&L responsibility and mid-term success measures allow the needed degree of freedom to conduct the innovation project. The organizational positioning under a strong and financially well-equipped innovation board is getting more important when the entity is predominantly customer project driven, the innovation is going to influence existing portfolio elements to a great extent, high amounts of personnel resources are needed and in markets where it is more difficult to acquire pilot customers, because of the need to invest big budgets. The presence of pilot customer and/or an innovation partner does not affect the necessity of process support for resource allocation and innovativeness, and hence, does not influence the decision to conduct the project under a supportive structure within the company's boundaries.

As soon as the solution reaches its commercialization phase, a transition into the line organization and alignment of sales systems and activities is mandatory. The more radical and disruptive the new solution is, the more a strong innovation board is needed to persuade senior-, line-, and project portfolio managers to foster it as a new portfolio element (Beringer et al. 2012). For the project team the transition of the solution or the integration of the project structure in the line implies a decrease of their degrees of freedom, because various people are dependent on reliable statements, e.g., the sales team needs information if and when a specific function is available, otherwise, work and cost intensive manual adjustments might be necessary. As long as the team works in the project modus, it can focus completely on the project tasks.

7.2.2 Companies processes

Coming from a creative idea to a successful marketable solution, a quick time to market possibility is indispensable, hence, companies' processes need to be in line with the described practical process model (cf. chapter 7.1.1; cf. figure 15). In contrast to the majority of incumbents and corporations that have extensive processes focusing their existing portfolio elements, processes need allow, guide and support the conduction of disruptive innovation projects and act as a means to start communication, rather than bearing bureaucratic procedures.

Within the first steps, from ideation to the set-up of an innovation project, especially in customer project driven company's, managers need incentives to allow serendipity and conduct first assessments of emerging disruptive ideas (Grote et al. 2012). This may be in terms of ex ante agreements and/or ex post recognitions. It must not be negatively correlated to the managers and employees target agreement, and hence, its personal compensation, consequently, either it is necessary to adapt them or to allow charging the costs to a corporate account.

Similar challenges occur when the allocation of personnel resources and shared services is necessary while conducting the innovation project. Traditionally, boards and processes need to be passed, requiring extensive formalities and demand the preparation of documents to start a fixed procedure starting at fixed dates. This results in time consuming bureaucratic work and idle times, what is inappropriate for innovation projects and constrains them. Moreover, traditional processes are not designed and laid out for cross-industry innovation projects what leads to the situation that the projects have to compete with ones related to the core portfolio for scarce resources, e.g., in terms of short-term financial indicators (Bond & Houston 2003). It is not just possible to avoid the existing processes, because otherwise, some required resources might be blocked through other projects or the process owners are not willing to provide the resources. Moreover, the team cannot buy them externally without following other extensive and time consuming processes because of corporate governance principles. As a solution and to foster leaner and quicker processes, to avoid waiting time for experts and/or shared services and to avoid rejections due to inadequate evaluation criteria, the required processes need to be adapted and fast-tracks need to be established. Due to the fact that process owners are generally not willing to treat any project special to avoid decreasing the importance of the process and/or respective boards, the innovation board needs to intervene and discuss potential solutions with the process owners. This includes among others new evaluation criteria with mid-term-based assessments, standardized templates among synchronized processes, assigning one contact person in each supporting entity who deals with the entities specific requirements, and especially no comparison of short-term revenue forecast with incremental innovation projects of existing portfolio elements (Bond & Houston 2003). Top-management hierarchy and/or support of the innovation board support the discussions, a redefinition of the resource allocation process and a potential reassessment of the process owners' personal incentive system.

Beside the internal processes, also processes related to contracts with external parties need to be aligned to the speed-boat approach. This includes among others lean contracts with innovation partners, pilot customers and external experts. Contracts related to the cash cows of the company are usually very extensive and cover various contingency. These are important when dealing with the company's portfolio elements with high revenue volumes, but for innovation projects it is necessary to have lean specified contracts that can be closed within adequate time and without extensive juridical consultancy.

As soon as the first evaluations have been conducted the innovation board meets on short notice, potentially decides to set-up the innovation project, and consequently, provides a pre-budget for further evaluation. The decision is predominantly based on the earlier design concepts, externally gathered information and business case documents, and the project set-up is realized as soon as possible (Riel et al. 2011). From that time, the core innovation team, whose personal compensation is not related to any line products or projects, has to report any deviation or particularities to the innovation board, to guarantee that the project is still in compliance with the desired goal. As soon as the core innovation team has finished its detailed analysis and first discussions with potential partners or pilot customers, the innovation board reassesses the planned solution and releases a substantial mid-term innovation budget. Following a mid-term time frame is a management commitment to develop the planned solution and increases the degree of freedom for the innovation team, because it allows keeping reporting formalities to a minimum. In that early stage the outcome and resource planning of the innovation project are still uncertain, and hence, innovation project team and innovation board have the possibility to adapt it as required over time. For this reason continuous reporting to the innovation board is essential to intensify communication. However, its frequency and intensity follows more specific requirements than a predefined schedule and formalities, and its focus is to assure that the development is on track and not to apply for further budget or justify its continuance.

There is no need for the innovation board to justify any budget allotment, however, while its initiation in accordance with the company's strategy, a rough dimension of the investment volume has been defined and the sensuousness of an innovation board as a strategic instrument will be discussed and reviewed with a mid-term perspective (Floricel & Miller 2003). Besides positioning the innovation board as an independent entity within the company's boundaries, one main precondition is that the personal compensation systems of the members of the innovation board are aligned to the successful transfer of cross-industry innovations into the company's portfolio and not towards fostering existing cash cows (Bond & Houston 2003). To foster the anchoring of the new solution within the company's portfolio, similar challenges as in the resource allocation steps arise. Beside the necessity to adapt existing systems, the sales force need to be motivated to start sales efforts. Usually target agreements are done in the beginning of the year, and sales people focus on specific existing portfolio elements to increase their personal compensation. Hence, processes need to be established to change them during the year and/or to stimulate additional sales activities due to financial incentives.

7.3   Collaboration Fit

Other major principals that support the core category 'project', and hence, the derived practical process model (cf. chapter 7.1.1; cf. figure 15), have been iteratively discovered in the 'collaboration' category. In addition to the earlier described theoretical model that derived from the company category, the theoretical model derived from the collaboration category focusses on human and its collaboration, both with an organizational and personal view.

7.3.1 Mindset and Organizational Conditions

Innovation projects need innovative people; People with an innovation oriented, open mind, start-up mentality, creativity, motivation to innovate, and people that are able to work in dynamic, unstable environments and follow pragmatic approaches. Cross-industry innovation projects additionally need people who are motivated to take the challenge to collaborate with people from different industries, such as the requirement to learn new working habits and nomenclature, accept different viewpoints, and follow new processes and innovation approaches (Bach & Whitehill 2008). Hence, especially the core team of the innovation project and the experts who collaborate intensively with the partners should not be chosen because they are part of the entity where the new topic evolved or because they already have proper know-how in the field, but they should be handpicked out of the whole company's workforce, to fulfill the assigned roles (Rese et al. 2013; Mansfeld et al. 2010; Riel & Lievens 2004). Specific new industry expertise comes along with the maturity of the project and either grows internally, or by hiring external experts as soon as it is foreseeable that the new topic consolidates and is going to expand the company's portfolio. Changing the mind-set of people and the culture of whole entities is much more difficult and takes much longer, hence, the innovation boards top priority should be to identify suitable people for the project (Newman 2009).

Additionally, it is important that the team members are not bound in operative business, to avoid contradicting incentives regarding personal compensation related to existing business and development of disruptive innovation (Grote et al. 2012). They should be able to act autonomous and completely detached from the existing portfolio to increase its degrees of freedom to work towards pre-defined goals, namely the development of a sustainable business model and an innovative solution (Mansfeld et al. 2010). This also includes an assured mid-term budget approval, distinct executive management support, commitment across all hierarchies and minimal reporting processes to establish an entrepreneurial culture (Newman 2009; Hansen & Bunn 2009).

7.3.2 Common Preconditions and Project Structure

After the team identified potential partner organizations, they need to sound if they fit. The partner finding process should not be influenced or defined by economic or personal interests, such as major customers or friendship of individuals (Grote et al. 2012). The team should solely focus on the planned innovation to identify a partner with complementary expertise and no conflicting interests. It is essential that the partners invest high efforts in this sounding process and discuss and clarify the goal of the collaboration, both in terms of the technical solution and in terms of the common business model. A common consensus is necessary what everybody brings in and get out, and it must be clear how everybody participates at the realized solution to reach a win-win situation (Gillier et al. 2012; Weck 2006). For this, it is required that all involved parties need to understand their own business model and the one of the partner. When it later turns out that the partners have similar egoistic commercialization interests, affecting same parts of the value network, trust and willingness to open collaboration inevitably gets negatively affected and everybody tries to expand its part in the value chain, what constricts the further development to a great extent. Roles, tasks and interfaces need to be defined and fixed in a project plan, project risk planning and goals need to be quantified and collaboration contracts need to be concluded that cover resource input within the development phase and also consider commercialization and property rights (Salomo et al. 2007; Weck 2006). Collaboration in form of a strategic exclusive or non-exclusive partnership, or under a formal juridical structure is possible. The selected structure is important for the collaboration with the partner, but does not affect the internal innovation project that is conducted and steered detached within the company's boundaries.

Setting up the common project also includes conclusions on project management, project administration and conflict resolution. A common steering board across the project structure is an effective means to reach mutual conclusions and to solve disagreements across organizations. It is advisable that beside the project manager(s), representatives from the innovation board are part of the steering board to guarantee that their personal incentives are detached from the existing portfolio, and hence, they thoroughly focus on the innovation project. The steering board needs to meet regularly to monitor the progress of the project and to potentially agree adjustment measures. Due to completely different innovation approaches, speed, and working habits, generally prevalent in different industries, the steering board also governs the synchronization of processes and that every partner follows its major processes, because otherwise, they subordinate too much and lose their efficiency and identity, revealed in various attempts where corporates integrated start-ups that gradually lost their innovativeness and dynamics. Reporting processes with the steering board and among the project team should be established as a means to keep communication and mutual exchange on a high level, but should not include excessive preparation and documentation. The availability of collaboration tools as integration mechanisms, encourage and intensify communication, support the collaboration (Grote et al. 2012; Kale & Singh 2009). Beside e-mail, telephone conferencing and project management tools, especially common data rooms and technical collaboration environments should be provided as early as possible, even though they might be new for some members of specific companies or industries.

7.3.3 Harmonization and Intensity of Collaboration

Cross-industry innovation projects usually involve that people with different backgrounds, working habits and innovation approaches meet each other. The differences are much more distinct compared to open innovation projects where companies from the same industry collaborate. Industry specifics every time have to be kept in mind when collaborating with the partners because they define the way people act. Social factors have a much higher impact compared to the technological factors, and a fit on the personal level generally implicates a fit on the business level as well. Hence, especially in the beginning of the collaboration it is important that the team members are open and motivated to get involved with each other and invest a lot of time and energy in familiarization (Grote et al. 2012; Granovetter 2005; Hoegl et al. 2004). In this phase the team agrees on a common nomenclature, learns the basics of industry specifics and establishes mutual understanding and a trustful environment. Throughout the project openness and tolerance and the acceptance of other working habits and views is important to support this trustful atmosphere, what is the basis for exchanging industry specific expertise that is important for the service creation (Getha-Taylor 2012; Hirst & Mann 2004). The willingness to collaborate is more important than defined processes and steering, but the project management need to facilitate and foster an intensive communication and collaboration to keep everybody informed and, potentially also in a non-business environment, to form a common cross-organizational team (Schreiner et al. 2009; Ritala et al. 2009).

The intensity of collaboration across the project phases is project and industry specific. In general, collaboration with partners is much more intensive within the customer project than in the innovation project, because the innovation project has strategic relevance and touches deep in the company's processes. Whenever the borders between innovation and customer project blur, there are still activities that are strictly kept inside the company such as internal business case calculation, strategic considerations and internal planning and communications. The activities of ideation, market analysis and rough design are rather done separately as a first instance, with less collaboration because everybody needs to be clear about the topic and form its own opinion regarding financial potential. It might be possible, that representatives from different companies develop the initial idea and pass the first steps of analysis together, but each company still needs to conduct its own internal assessment. In the course of the project the intensity of collaboration increases steadily, from first sounding discussions to a very intensive initialization phase, that includes social team harmonization, groundwork, industry know-how transfer and the definition of working packages. Subsequently the collaboration intensity gets a bit less, because work need to be done in the working groups, but, supported by the project management and collaboration tools, it is still on a very high level to keep everybody informed, allow industry specific knowledge exchange and because of reconciliation works. In the end of the development the collaboration intensity increases again, to bring the working packages together and to launch the commercialization activities that can be either a common or a separate activity.

7.4 Portfolio Expansion

Pooled in the 'market' category, various concepts supporting the core category 'project' have been identified in the analysis phase. This chapter extends the derived theory with market based portfolio considerations, predominantly related to fundamental strategic decisions, such as the strategic positioning in the market.

7.4.1 Transfer and Anchor of Innovation

Portfolio management allows senior management to implement the pursued corporate strategy (Unger et al. 2012). Project developments not in line with the strategy need to be terminated, while desired innovation reaching a certain maturity and marketability, need to be transferred from a project mode into the line organization – given the nature of an innovation project that gradually ends when the solution reaches the state of a new product. Usually telecommunication companies have processes in place that govern the transfer of new features of existing products and cover topics such as juridical assessment, adaptation of procurement systems, marketing material and training of sales force. The transfer of completely new services and disruptive innovation however, poses bigger challenges, because they are the spearheads of a structural change, that may also cannibalize the current revenue stream and that typically reach lower revenues in the short term, what negatively influences the existing personal incentive and compensation system (Chesbrough 2010, pp.94–96; Bond & Houston 2003). Hence, it is difficult to find internal business owner who support these new services and they even face blockades in the organization that has typically not yet adapted for the change. Cutting these organizational barriers of conflicting interests is one of the major challenges for the innovation team and especially for the innovation board, to allow an unbiased discussion of a most suitable transfer. This also includes the approach of various internal stakeholders to get juridical approvals, to adapt systems and documentation, reach sales commitments, and to either train current employees in industry specific knowledge or to consider recruiting new employees with the relevant industry knowledge, what makes the company more independent from further partnerships and strengthens its position in the new market. These actions take time what contradicts with the requirement that innovation typically need a quick time to market. Hence, the innovation team should pave the way for a smooth transition as soon as possible, initiate discussions and prepare required documents. The external communication however, needs to be planned very carefully. Market announcements that do not get fulfilled or that are available with a long delay decrease the reputation in the new market.

7.4.2 Self-Concept, Strategy and New Services

Cross-industry innovation projects either result in innovation for an *existing industry* or for an *emerging market* due to converging industries. The former comes with the decision to offer the innovation for the market and act as an enabler or to offer the innovation in the market and take a role in the market that generally implicates the role of a competitor. The situation in emerging markets is different, while the differentiation also applies, because typically the market is more dynamic and roles are evolving while various players gradually enter the market. A first classification how to approach the market is the basis for the internal business case when assessing the projects profitability. This need to be in line with the corporate strategy and verified while conducting the project and especially before implementing the solution in the portfolio (Bond & Houston 2003). Even though some innovation projects are possible without innovation partners, and hence, with a wide scope of action, it is an alternative to stay inside its self-imposed boundaries, not pursuing an active role in a specific market, while voluntarily resigning from parts of the value chain.

Typically the solution that arises from the first cross-industry innovation project allows developing various side products that complement the initial solution. To develop a whole ecosystem of complementing services, the telecommunication company is compelled to think outside the boundaries of its own initial value creation (Chesbrough 2010, pp.107–109; Miller & Olleros 2007). It has to understand the whole industry context, and not just the part of its own contribution, to decide what to contribute in addition. Service innovation is directly related to business models that support these services and services can only be successful in the long run with a viable business model that creates value for its customers and providers (Chen 2011; Chesbrough 2010, pp.89–111). As a form of extended business development, conducting customer workshops is a good means to evaluate business models and to identify what kind of business cases the companies in the industry are able to realize based on the new solution. With every emerging idea the process model starts over, assessing the idea and developing and commercializing the solution with a suitable partner organization (cf. chapter 7.1).

Cross-industry innovation projects are getting more and more popular throughout all industries and telecommunication companies have the prerequisites to conduct them. First of all they have ICT know-how and fast innovation cycles that allow a fast go-to-market. Hence, telecommunication companies should consider not just to extend their portfolio with solutions they develop in cross-industry innovation projects, but also to support other companies while conducting these kinds of projects and act as innovation accelerator (Howells 2006). They can easily realize a supportive role in form of a new portfolio element that offers various modules to support temporary cross-industry innovation projects in terms of consulting and ICT provisioning. Consulting services focus on how to successfully conduct cross-industry innovation projects, especially process guidance, industry specific expertise, organizational structure and applying fast innovation cycles. While ICT provisioning consist out of a collection of collaboration and communication tools for content management/common data rooms, telephone/web conferencing and/or telepresence, common development/technical collaboration environments and project management (Plewa et al. 2012)[4].

7.5 Overall Considerations in Practical Realization

In addition to the implications around the categories project, company, collaboration and market (cf. chapter 7.1 – 7.4), this chapter specifies general thoughts and recommendations, for telecommunication companies and for potential partner companies, that need to be considered while applying the emerged theory in practice.

7.5.1 Impact and Interdependency within the Telco Organization

Cooper & Edgett (2012) note that nearly three-quarter of participants in their study have some kind of formal new product development process in place, and that 90 percent of best performers, compared to 44.4 percent of worst performers, have a clear, defined new product development process, that guides new product development projects from idea to launch. Best performers are

---

[4] Plewa et al. (2012) note that perceived usefulness and compatibility of innovation management applications are the key for adaption, and hence, innovation process performance. The authors suggest that a number of activities can be employed to enhance user perceptions regarding perceived usefulness, e.g., training.

between two and three times more likely to have implemented a successful new product development process than worst performers (Cooper & Edgett 2012). Moreover, performance is boosted if the company follows a staged new service development process (Song et al. 2009). This suggests that organizations should reflect the proposed practical process model (cf. chapter 7.1 – 7.4), adapt their innovation processes and anchor it within the whole organization, to consider specifics of cross-industry innovation projects (Chapas et al. 2010; Grönlund et al. 2010). The proposed practical process model is one possibility for telecommunication companies to increase the probability of performing successful cross-industry innovation projects. For all non-telco companies it presents an optimal basis to describe their own process models that may be similar to a great extent (cf. chapter 9.3).

The growing number of projects sharing and competing for scarce resources require comprehensive project portfolio management, for a strategic alignment and efficient use of resources among the projects (Voss 2012; Beringer et al. 2012). One of the major challenges of implementing the proposed model will be the adaption of existing processes. An organizational change process, to implement open innovation in practice, comprises the stages of unfreezing, moving and institutionalizing and helps organizations to adapt inter-organizational networks, organizational structures, evaluation processes and knowledge management systems (Chiaroni et al. 2011; Chiaroni et al. 2010). The implementation can be done in the most effective way when the innovation board has the required strength and reputation. This can be forced by the top management board to a certain degree, because management support sets the course for its strategic direction (Floricel & Miller 2003). Moreover, an R&D entity can strongly increase its influence when it achieves a high degree of innovativeness in its activities and when it is highly connected to the customers (Engelen & Brettel 2012). The degree of innovativeness generates dependency on the innovation entity, while the connection to the customers demonstrates that the outcomes are at least partially in line with the customers' needs. Both increase the availability of needed resources.

However, a clear separation of the innovation project from the operational business is essential for its success. The degree of integration with other departments negatively impacts the degree of R&D influence (Engelen & Brettel 2012). E.g., managers of the telecommunication company opined that the speed-boat mentality was less intense in the strategic areas compared to real start-ups, because they were embedded in the telco's B2B entity that had its own innovation management that mainly focused on B2B-ICT solutions, influenced by a customer project driven approach. It was very seldom that the company sold earlier developed generic solutions. Moreover, they were bound in its organizational processes and had short to medium term financial goals. This supports the suggested organizational precondition (cf. chapter 7.2) that the innovation team should be able to independently command resources and that the personal resources should be free of conflicting incentives. Employees need specific degrees of freedom and an open mindset to conduct comprehensive cross-industry assessments (Mansfeld et al. 2010). Organizations that lack that capacity should consider a partner to manage both the outside-in analysis and the development of external sources and ideas (Nolf et al. 2012).

Gales & Mansour-Cole (1995) raise the question if long-term high quality collaboration relationships support innovation projects, or if these tight ties rather are disadvantageous (Gales & Mansour-Cole 1995). There is a trade-off between the opportunity of innovativeness and the risk of misunderstanding, grounded in the inverted u-shaped effect of cognitive distance on innovation performance (Nooteboom et al. 2007) (cf. chapter 6.3.1). Hence, there is always potential for disagreements and misunderstanding between partners from distant industries. Long-term, established relationships help to overcome difficulties to anticipate market needs, may reduce interference and disadvantageous coordination, but it may also cause rigidities and accordingly less innovativeness (Gassmann, Zeschky, et al. 2010; Eisingerich & Bell 2008; Foxall & Johnston 1987; Terreberry 1968). Rather the partners should revert to strategic design tools that simultaneously support managing and coordinating the diversity of interests between the partners and that support sustainably scheduling and planning of resources and skills (Gillier et al. 2012; Gillier et al. 2010). Additionally they should reassess the partner selection process whenever necessary, especially at the initiation of each new project.

7.5.2 Transferability – Deliberations for Potential Partners

Network management approaches should not be generalized across industries or sectors, because their differences matter (Buganza et al. 2011; Herranz 2006).

Nevertheless, the topic of cross-industry innovation is not just relevant for the ICT industry. Also, in other high-technology industries, such as pharmaceutical, healthcare or aerospace/defense, companies need to motivate their employees to look not just beyond group or divisional lines, but beyond corporate and industry boundaries to identify and develop innovative solutions for present-days challenges (Fraser et al. 2007). Low-tech industries, such as machinery, turbines, medical tools, fast moving consumer goods, food, architecture or logistics lag behind the high-technology industries. There are first observations that they are opening up their innovation processes, to exploit open innovation potentials. E.g. in the early phase of innovation they involve users systematically and they start to open up in all other directions as well (Gassmann, Enkel, et al. 2010). Overall, the more a company operates in a high-technology industry, and the more the activity can eventually become core business, the more it is involved in strategic technology alliances. While the more a company operates in a low-technology industry, and the more the activity is their core business, the more M&A is a prevalent form of integration of external sources of innovation (Hagedoorn & Duysters 2002).

Gassmann et al. (2010) observed a bandwagon effect in their executive education programs; "CTOs with closed innovation models and strong internal R&D are under increasing pressure to justify their refusal to cooperate with the outside world and exploit the open innovation wave." (Gassmann, Enkel, et al. 2010) Open innovation spreads across industries, from high to low tech, from large firms to SMEs and from pioneers to mainstream (Gassmann, Enkel, et al. 2010). However, open innovation is not per se better than a traditional more closed approach, instead the intensity and quality of the relationships determine innovation outcomes; projects with a high level of novelty demand a high level of communication and knowledge sharing (Hsieh & Tidd 2012). Consequently, open innovation and in specific cross-industry approaches are more relevant for more radical innovation projects – across all industries.

8 Impact on Existing Theoretical Innovation Process Models

Following Österle et al. (1991, p. 35) and Riempp (2004, pp. 314-316) a profound research process not just provides practical, but also theoretical contributions (Österle et al. 1991, p.35; Riempp 2004, pp.314–316) (cf. chapter 1.3). Putting the explicitly described practical strategic implications for telecommunication companies (cf. chapter 7) in reference with existing innovation models (cf. chapter 2.3) it gets obvious, that major characteristics of cross-industry service innovation projects are not considered in present literature to its full extent. Accordingly, based on the new insights this chapter deduces cross-industry innovation specific propositions and visualizes them in an extended theoretical innovation process model.

8.1 Requirements for Theoretical Innovation Models

No innovation process model can fulfill the requirements of generality, simplicity and accuracy at the same time. Network models may show main point of contacts and linear models may consist of a few fairly generic stages such as fuzzy front end, new product development and commercialization, and hence, may be generally applied to various situations but they do not provide detailed insights (Gassmann, Kausch, et al. 2010). In contrary, the models may be very accurate and reproduce a whole ecosystem of internal and external linkages, and sequential models may consist of around ten stages that may be applied to a very specific setting (Alam 2007; Alam & Perry 2002). Consequently, there is not one correct model, but a wealth of models depending on the respective situation and circumstances, and all existing innovation process models have a right to exist (Harmancioglu et al. 2007).

Horizontal collaboration within the same industry is not a new approach. Increased cost and uncertainty regarding R&D efforts, especially in high-technology industries, have forced organizations to establish partnerships and various models covering strategic alliances exist (Teece 1989; Nesse 2008; Trott & Hartmann 2009). These partnerships predominantly appear within one industry, with a closed innovation process approach. Horizontal open innovation across industry boundaries is still a new topic, both in literature and practice (Gassmann, Enkel, et al. 2010), hence, just a few innovation models exist, that either do not cover ICT industry specifics or are too general and simple for extensive insights. Moreover, most recent open innovation models focus on vertical relationships and describe the specifics of open innovation with customers or lead user integration, or supplier as innovation partner. Consequently, no precise innovation process model exists, that covers the specifics of cross-industry open innovation in the ICT industry. This suggests that existing models need to be expanded and adapted to the present study focal research area.

In addition, the present study focusses on services in terms of non-physical, intangible products. Despite the practical importance of the service sector, today's innovation research is mainly product oriented and rarely includes service innovations, compared to physical product innovations (Thomke 2003; Yang 2007; Nesse 2008; Gassmann, Enkel, et al. 2010). Most innovation models focus on pure goods and a few on pure services, but not on the characteristics of intangible products (Segelod & Jordan 2004). The present study shows that ICT-services focused in the case studies (cf. chapter 4) cannot be treated in the same way like pure goods or pure services, and hence, ICT-service specifics need to be considered as well, while adapting existing innovation process models.

8.2   Propositions for Theoretical Innovation Process Models

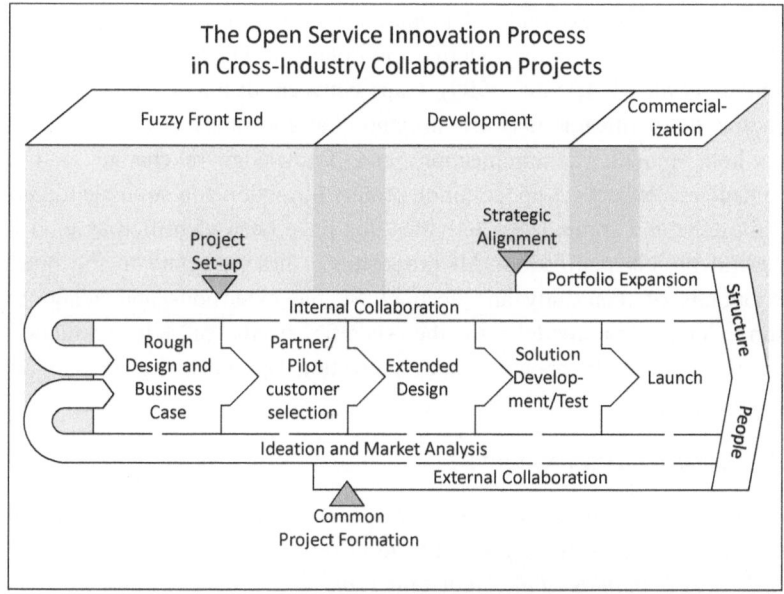

Figure 16: Extended Theoretical Innovation Process Model

Figure 16 visualizes the deduced propositions, embedded in existing innovation process models. In the upper part, a general and simple innovation process consisting of three consecutive phases, namely fuzzy front end, (new product) development and commercialization defines the basic framework (Gassmann, Kausch, et al. 2010). It directs ones attention to the relatively long early innovation phase. In the lower part, the figure combines various four to five stage model characteristics and accumulates parallel activities as well as further innovation network dimensions. The phases ideation and market analysis, rough design and business case, partner / pilot customer selection, extended design, solution development/test and launch originate from the practical implications (cf. chapter 7) and can be partly determined in existing process models as well (Eveleens 2010; Song et al. 2009; Gales & Mansour-Cole 1995). It needs to emphasize that ideation and market analysis are continuous parallel activities and that the phase of partner / pilot customer selection has a distinct position in the process. Other network dimensions encompass and extent the phases. People, structure and strategy (portfolio expansion) have been defined as dimensions interweaved and linked with the phases (Roberts 2007). Following characteristic thoughts of the fourth and fifth generation innovation process models internal and external collaboration are parallel activities and integrate internal networking and the external environment (Du Preez & Louw 2008; Blomqvist et al. 2004; Niosi 1999). Gaps between all major connecting points symbolize the permeable network structure that allows a forth and back across the whole model, communication and knowledge exchange and also dependencies. Project set-up, common project formation and strategic alignment are milestone and compulsory activities that have been identified as vital for a successful project realization. All propositions directly relate to the practical contributions of this study and account for the extensions and adaptions of existing innovation models to the specific needs of telecommunication companies and their open service innovation approach for cross-industry innovation projects.

8.2.1 Extended Fuzzy Front End

Proposition 1: The long duration of the early innovation phase needs to be considered in the process to guarantee realistic planning and time management from the very beginning of the innovation project.

One of the simplest open innovation processes, consist out of the three consecutive phases, early innovation phase / fuzzy front end, technical product development / new product development and commercialization (Gassmann, Kausch, et al. 2010). The process is simple, generic and may explain most innovation approaches in organizations across industries. However, to gain valuable insights and recommendations, it needs a deeper analysis of the three phases. Gassmann, Kausch & Enkel (2010) suggest partly parallel and dependent sub-phases: concept definition, idea & functionality selection, idea & knowledge creation, opportunity identification & analysis and prototype testing (Gassmann, Kausch, et al. 2010). In the present study various similar activities that consist analogical purposes belong to the early innovation phase: ideation and market analysis, rough design and business case, internal project set-up, partner / pilot customer selection, collaboration with internal entities and first external collaboration, common project formation and to a certain degree an initial common extended design. No matter what exact activities are considered, it is obvious that various resource and time consuming activities need to be considered, what – at first sight – conflicts with the fact that in practice innovation need a quick go-to-market. Hence, organizations tend to neglect the early innovation phase, to start very early with the subsequent development phase to present early achievements. But even though these activities happen before the actual development, they nevertheless, are of utmost importance to guarantee a successful realization of the innovation project as a whole, and underline the importance of the early innovation phase (Grote et al. 2012). Hence, they need to be integrated and emphasized in academic models as well as in organizational planning. In some cases it may be possible to streamline the sequence of the activities and where possible to follow a simultaneous approach, but nevertheless, the early innovation phase / fuzzy front end is a time consuming period that should not be underestimated or consciously reduced (cf. figure 16).

8.2.2 New Explicit Phase in the Innovation Process

Proposition 2: The importance of selecting suitable innovation partner and/or pilot customer is inevitable for the success of the project and hence, needs to be integrated as a distinct new phase in the innovation process.

As soon as the company has completed its internal assessments with its given internal capabilities, it needs to decide if the planned solution can be specified and bought at the market, an internal self-development is possible and desired, or if a partner company is needed. The later generally applies in cross-industry innovation projects, when comprehensive industry-specific knowledge is not available in the company boundaries. It requires a thoroughly market screening for a potential innovation partner that is willing to bring in the required industry specific knowledge and guarantees a strategic fit regarding potentially common commercialization activities. The qualitative analysis in the present study (cf. chapter 4 and 5) makes clear, that in cross-industry innovation projects, the success of the whole project is inevitably linked with the innovation partner and its market acceptance with the availability of a pilot customer. As described in chapter 5.2.1, the roles of innovation partner and pilot customer can be fulfilled by one company, that acts as pilot customer and innovation partner at the same time, or by independent organizations. Both need to be carefully identified, selected and integrated in an early stage of the innovation process to guarantee an early involvement and a mutual commitment.

In existing innovation process models the activity of partner selection is typically not considered as a distinct stage. Most models consider a phase of idea generation or searching for ideas for innovation as a starting point, a subsequent selection of which projects are pursued and then, immediately start with the development and testing phase (Eveleens 2010). Cross-industry innovation projects in contrast are dependent to a much greater extent on the partner organizations than described in traditional models. Partners contribute with new insights, allow reflecting the first internal assessments and results of ideation and market analysis, rough design and the first internal business case. They increase market acceptance, determine the structure of the innovation project, allow developing a common extended design and are required for the development of the desired solution. Partner selection and the subsequent definition of the common solution in an extended design phase can be regarded as the beginning of external collaboration and a fit of the organizations is of utmost importance, both regarding collaboration habits and strategic intentions (Bach & Whitehill 2008; Bhattacharya & Krishnan 1998). The selection process apparently is time and resource consuming when performed thoroughly, but the specifics of cross-industry innovation projects emphasize the importance of partner finding, imply to focus on the selection phase and strongly propose to implement it as an additional distinct phase in existing innovation process models (cf. figure 16).

8.2.3 Sequential Phases and Parallel Activities

Proposition 3a: Market analysis and ideation correlate and need to be permanent activities, both in creative innovative organizations and in its corresponding innovation process model.

Proposition 3b: Internal and external collaboration as parallel activities need to be integrated in innovation process models to visualize required internal networking and to link the process with the external environment.

A linear process model from ideation and market analysis to the launch of the solution, very similar to Cooper's stage-gate-model, is the core of the suggested model adaptations (Cooper 1990). It can be easily applied in organizations and includes the phases of: ideation and market analysis, rough design and business case, partner / pilot customer selection, extended design, solution development/test and launch (cf. chapter 7). Stage models chronologically describe the development of a project, its activities and challenges in a series of stages, however, they do not capture the fact that some activities take place simultaneously (Weck 2006; Burgelman 1983). Consequently, the main difference to Cooper's model is visualized by arrows, that are encompassed by the next phase to describe a fluent passage from one phase to another (cf. figure 16). Moreover, ideation and market analysis as the first phase is not placed as a clear starting phase of the process as typically in other innovation process models (Eveleens 2010). Rather it follows the observed facts (cf. chapter 4) that in particular in high-technology industry organizations market analysis is a permanent activity what implies the generation of innovative ideas. Hence, organizations should have processes and innovation tools in place to foster ideation, and the acquisition and dissemination of knowledge at any time (Jiménez-Zarco et al. 2011).

These processes and innovation tools also support the collaborative development of innovation across the entire organization (Jiménez-Zarco et al. 2011; Koppinen et al. 2010). Guided by the fourth and fifth generation of innovation models (cf. chapter 2.3) it is recommended to integrate internal and external collaboration activities as parallel activities along the linear core process model. The fourth generation of innovation process models emphasize cross-functional, parallel and interactive integration of innovation activities within organizations, and also stress horizontal strategic alliances, while the fifth generation models focus on systems integration and networking to realize a fully integrated parallel development, including strong linkages with leading edge customers and strategic integration of primary suppliers (Du Preez & Louw 2008; Nobelius 2004; DeSanctis et al. 2002; Niosi 1999; Iansiti & West 1997; Rothwell 1992). Considering internal and external collaboration as parallel activities emphasizes the fact that collaboration in cross-industry innovation projects is inevitable for a successful realization (cf. chapter 5). In the beginning of the innovation process, the core innovation team primarily works with internal experts regarding legal and technical matters, to refine the business case, to plan the realization and to conclude contracts, while in the later innovation phases, internal collaboration regarding the adaptation of processes and portfolio expansion come along (Blomqvist et al. 2004; Bond & Houston 2003). Especially for commercialization activities the team needs to be attuned with various entities across the organization. Collaboration with the external partners start within the selection phase, gets intensified while developing the common extended design to get a better understanding of the common solution, to plan responsibilities, the time frame and to set-up the project management. It may be possible to define working packages with decreased collaboration necessity, but communication and knowledge exchange needs to stays at high level and collaboration increases significantly when common commercialization activities are planned. Overall, grounded by findings in the cross-case analysis (cf. chapter 5) it seems appropriate to merge parallel and network characteristics of earlier generations of innovation models with a commonly practical applied linear model, what suggests in particular, to integrate internal and external collaboration as parallel activities in present innovation process models (cf. figure 16).

8.2.4 Innovation Network – Additional Components

Proposition 4: The dimensions of people, structure and strategy (portfolio expansion) need to be interweaved and linked with the phases, to symbolize that following a sequence of steps is not sufficient to successfully integrate an innovation process.

It can be considered as a shortcoming of present open innovation models that they are inherently linear, and basically a variation of the well-known stage-gate model (Trott & Hartmann 2009; Cooper 1990). At various parts in the present study, it likewise, got obvious, that a myopic focus on the detached innovation process, does not guarantee a successful realization of cross-industry innovation projects (cf. chapter 4 and 5), rather the process needs to be embedded in a complex ecosystem (Chapas et al. 2010). Cross-industry innovation projects require intense collaboration, hence, people need to be open for this new endeavor, to learn and accept industry specific working habits and point of views. If people do not accept its cross-industry partners or try to follow familiar procedures, it will not be possible to successfully realize and integrate the planned solution (cf. chapter 7.3). In addition, in the present study, the innovation projects were embedded in the organizational structure of the telecommunication company, hence, process specifics, especially regarding board requirements applied. Dependency on resources of the organization got obvious when expert support and budget approval was needed, or when systems had to be adapted for commercialization activities, because incentive systems were aligned to support the present portfolio (cf. chapter 7.2). Moreover, the organizations strategy needs to be considered when selecting an innovative project idea, to develop a sustainable integrated solution. Just after the idea generation phase, most authors consider a step to narrow the potential projects down and select an innovation project in line with the organizations strategy, its pursued role in the market and the existing portfolio of projects and products (Eveleens 2010; Howells 2006). Cross-industry innovation projects typically generate more radical innovation that may even cannibalize the existing product portfolio, what must explicitly be covered by top-management and divisions/departments heads decisions, and considered in the organizations corporate strategy (Beringer et al. 2012) (cf. chapter 7.4). Consequently, the core of the proposed innovation model adaptions consisting of a linear process, needs to be expanded by the dimensions of people/staffing, planning of organizational processes/structures/systems and plans/strategy (Yang 2007; Roberts 2007) (cf. figure 16). The additional components allow to embed the easily practical applicable linear process in a whole interlinked innovation network, that guarantees a thoroughly consideration of influencing internal and external factors (Blomqvist et al. 2004).

8.2.5 Innovation Network – Characteristics

Proposition 5a: Project set-up, common project formation and strategic alignment are milestone and compulsory activities that need to be considered in the process, emphasizing team formation, external collaboration and internal alignment.

Proposition 5b: Collaboration is an imperative in cross-industry innovation projects, hence, the innovation network structure needs to be permeable to allow extensive communication and knowledge exchange across the whole model, to cope with uncertainties and rigidities.

The proposed innovation model adaptations follow present study insights and considerations of earlier generations of innovation models. It highly emphasizes openness and collaboration across the network, with internal and external entities.

The linear process in the lower part follows Coopers process model but avoids its characteristic gates after each stage, that some authors criticize as shortcomings (Cooper 1990), furthermore there is a new trend in open innovation research, from stage-gate to probe-and-learn, hence, the stages should be permeable to allow an iterative and interactive process and explorative development (Gassmann, Enkel, et al. 2010; Weck 2006; Lynn et al. 1996). In the proposed model each stage goes over into the next, visualized by arrows, that are encompassed by the next stage, to describe a certain overlapping and a fluent transition from one stage to another. Hence, no strict gates like in the traditional stage-gate process exist, but the most characteristic points of project set-up, common project formation and strategic alignment, have been explicitly incorporated in the model (cf. figure 16). They shift Roberts' (2007) and Yang's (2007) dimensional thoughts of people, structure and strategy from an organization-wide view to the project level (Yang 2007; Roberts 2007). In addition they indicate that typically in cross-industry innovation projects an internal innovation project is accompanied by an external innovation project with an innovation partner and strategic considerations are major specifics.

The innovation process is not explicitly designed to go back from one stage to another, because the stages are self-contained, e.g., from the partner selection phase is no need to go back to the rough design phase, because in the next phase an extended design is developed with the partner company (Bhattacharya & Krishnan 1998). Also, development and test are incorporated in the same phase and the set-up of production infrastructure and physical distribution of material can be neglected because of intangible service characteristics.

Innovation models emphasizing cross-functional, parallel and interactive collaboration within an organization, networking and horizontal strategic alliances, allow access to a much larger base of ideas, knowledge and technologies (Du Preez & Louw 2008; Gassmann & Enkel 2006; Nobelius 2004; Gassmann & Enkel 2004; Iansiti & West 1997; Rothwell 1992). However, despite of its focus on knowledge integration, the models predominantly represent a closed innovation view, because the majority of the innovation activities happen within the company's boundaries. Most innovation studies do not recognize the value of partners outside the value chain to its full extent (Enkel & Gassmann 2010). In cross-industry innovation external input sources and extensive communication activities with the internal and external environment need to be considered across the entire network to a much greater extent, to foster knowledge accumulation and processing (Gassmann, Enkel, et al. 2010; Galanakis 2006). To illustrate this approach, open spots at boundaries and transition points of parallel activities and linear phases have been incorporated across the whole model (cf. figure 16). They visualize the permeable character of the model and implicate a dynamic, open environment that fosters knowledge exchange and collaboration.

9 Conclusion

This chapter summarizes the key findings of the present study and its contribution to contemporary theory and practice. It also identifies limitations and suggests topics for further research.

9.1 Practical Contributions

Innovation emerge when individuals and organizations start to look outside the box and break the traditional moulds of their own industry (Telstra Corporation Ltd. & KPMG International 2012; Song et al. 2011). These cross-industry innovation tend to be more innovative than the ones introduced within the own industry and cross-industry cooperations tend to introduce more radical innovation compared to organizations that are cooperating within the same industry (Kotabe & Swan 1995). As a result, cross-industry innovation is a strategic imperative for telecommunication companies to overcome present challenges in the ICT industry.

The present study's main objective was to describe how telecommunication companies should conduct cross-industry service innovation projects, by means of a practical process model serving as a guideline. It got obvious very early in the research process that an innovation process model alone does not guarantee the success of cross-industry innovation projects. Rather the organizational structure needs to be considered, people are from utmost importance and a strategical alignment is inevitable to guarantee portfolio expansion or other commercialization (Roberts 2007). Based on within-case analysis and a subsequent cross-case analysis (cf. chapter 4 and 5) it was possible to process first findings and to deduce preliminary implications. These have been reflected in the context of literature review (cf. chapter 6) to deduce practical strategic implications for telecommunication companies conducting cross-industry innovation projects (cf. chapter 7).

As the focal element, the practical process model illustrates which steps and decisions need to be considered (cf. chapter 7.1 and figure 15). It emphasizes the importance of a pilot customer for commercialization reason and the 'fit' and alignment with an innovation partner to increase the overall project success. The roles of pilot customer and innovation partner can be fulfilled by one partner or by several organizations. Moreover, the process model stresses the fact that a customer project may serve as a knowledge base in parallel to an innovation project, or it can form the nucleus of an innovation project. While the former is a means to prove the concept, the later allows a more cost-efficient development, and hence, is the preferred approach.

In addition, various parallel activities are essential, that link the linear process to a value network embedded in an organizational structure (cf. chapter 7.2). Due to the dependency on organizational shared service infrastructure and resources, and in order to quickly transfer the innovation into the existing product portfolio, the innovation project needs to be conducted within the company's boundaries. However, it should be separated from the operational business and should not need to compete with incremental innovation projects related to the core portfolio, for scarce resources, because extensive process and incentive systems are aligned to the core business (Beringer et al. 2012; Bond & Houston 2003). A strong and financially well-equipped innovation board consisting of determined top-managers whose personal compensations are not dependent on the core portfolio elements of the company, will help to establish leaner processes for the strategic areas across the organization (Floricel & Miller 2003).

In cross-industry innovation projects, literally people from two worlds met each other. They are influenced by different industry characteristics, such as nomenclature, working habits, innovation speed, -processes and –approaches (cf. chapter 7.3). This situation demands handpicked, open minded and tolerant experts out of the whole company's workforce, that are willing to understand and respect new approaches and thinking. These characteristics, increase the chance of a personal 'fit' with the innovation partners, what supports confidence, trust and harmonic working habits, as a basis to exchange needed industry specific expertise (Gassmann, Zeschky, et al. 2010). Intensive and transparent face-to-face communication, especially in the beginning of the collaboration also brings the partners closer together (Hoegl & Gemuenden 2001). A steering board and telephone/web conferencing and common data rooms need to be established to keep the mutual exchange on a high level throughout the project. Conflicting interests will inevitably negatively influence trust, the willingness to open collaborate, and hence, the success of the project, because each partner will attempt to expand its part of the value network. Well-defined formal processes and contracts, can never cover all contingencies, hence, to reach a win-win situation from the beginning throughout the project, the common business model needs to be understood and it needs to be determined what every party brings in and gets out (Weck 2006).

Incumbents typically struggle while commercializing radical innovation, because internal technologies, processes and incentive and compensation systems are aligned to the existing product portfolio, and they fear to open a Pandora's box that may decrease their influence (Hill & Rothaermel 2003; Bond & Houston 2003). Consequently, people hesitate to support a transfer of the developed solution and to expand the present product portfolio, until an internal radical innovation friendly environment is aligned to quick commercialization (cf. chapter 7.4). Moreover, the organization needs to decide which role it is going to pursue in the market, what may be the role of an enabler or competitor to established market players (Howells 2006). The decision needs to be in line with corporate strategy and based on business models that also consider that typically the solution that arises from the first cross-industry innovation project allows developing side products that complement the initial solution and may constitute a whole new ecosystem of complementing services (Newman 2009; Miller & Olleros 2007).

9.2 Theoretical Contributions

Although existing innovation process models do not entirely reflect the characteristics of cross-industry service innovation projects in the telecommunication industry, they are a useful basis to discuss its specifics (Ottenbacher & Harrington 2010). Furthermore, current research trends seize the topic and suggest to address it in future research (Gassmann, Enkel, et al. 2010). Therefore, guided by the research process of Österle et al. (1991, p. 35) and Eisenhardt (1989) the present study's objective is to provide new theoretical insights and propositions for existing open innovation process models, in addition to practical contributions (Österle et al. 1991, p.35; Eisenhardt 1989).

The practical strategic implications (cf. chapter 7) have been translated in demands for existing theoretical models in literature, theoretical implications in form of propositions have been deduced and visualized in an extended innovation process model (cf. chapter 8.2 and figure 16).

The study reveals that various important, resource and time consuming activities need to be considered in the fuzzy front end phase of the innovation project. This typically conflicts with organizations intention of a quick go-to-market, and hence, they tend to neglect or condense this phase. Consequently, a long duration of the early innovation phase needs to be considered in existing innovation process models, to guarantee realistic planning and time management from the very beginning of the innovation project (cf. chapter 8.2.1).

One of the activities in the early innovation phase is the selection of a suitable innovation partner and/or pilot customer. It is of utmost importance to find a partner that fits in respect to collaboration habits and strategic intention. This step typically is not considered in present innovation models, but it is inevitable for the success of a cross-industry innovation project, and hence, needs to be integrated as a distinct new phase in a more detailed innovation process (cf. chapter 8.2.2).

Most innovation process models start with ideation and market analysis as first steps. However, it needs to be emphasized that even when the project is in an advanced state new market insights and ideas do influence the further actions, and in most high-technology organizations they are permanent activities and not just in the beginning of an innovation project. Market analysis generates ideas, and ideas can be probed while doing market analysis. Consequently, market analysis and ideation correlate and need to be permanent activities, both in creative innovative organizations and in its corresponding innovation process model (cf. chapter 8.2.3). Moreover, by definition cross-industry innovation projects demand extensive collaboration across the whole project. Hence, guided by the fourth and fifth generation of innovation models, internal and external collaboration need to be integrated as parallel activities in innovation process models, to visualize required internal networking and to link the process with the external environment (Mortara et al. 2009, pp.30–40) (cf. chapter 8.2.3).

Linear innovation process models may cover all steps that are necessary to successfully conduct innovation projects in theory. However, the present study reveals at various points, that in practice supplementary factors influence the detached innovation process (cf. chapter 4 and 5). To increase its relevance additional components need to be considered that embed the process in a complex ecosystem. In cross-industry innovation projects collaboration is of utmost importance, because typically the partners are dependent on each other's industry-specific knowledge and expertise (cf. chapter 7.3). Moreover, the innovation project is embedded in an organizational environment that determines the availability of resources and process necessity (cf. chapter 7.2). In addition, most authors consider innovation project selection in line with the organizations strategy and the existing portfolio, what is even more relevant for radical innovation, because the innovation may not fit in the portfolio and/or the organizations role in the market needs to be defined (cf. chapter 7.4) (Eveleens 2010; Newman 2009). Consequently, the present study suggests that the dimensions of people, structure and strategy (portfolio expansion) need to be interweaved and linked with the phases, to symbolize that following a sequence of steps is not sufficient to successfully integrate an innovation process (cf. chapter 8.2.4).

The same considerations of Roberts' (2007) and Yang's (2007) dimensional thoughts of people, structure and strategy may be shifted from an organization-wide view to the project level (Roberts 2007; Yang 2007). This suggest that project set-up, common project formation and strategic alignment need to be considered in the process as milestones and compulsory activities, emphasizing team formation, external collaboration and internal alignment (cf. chapter 8.2.5). This proposition and the previous ones may be combined to an extended innovation process model (cf. chapter 8.2). In such a model, the whole network structure needs to be permeable to allow extensive communication and knowledge exchange across the whole model, to cope with uncertainties and rigidities (cf. chapter 8.2.5).

9.3 Limitations and Further Research

The present study contributes significantly to present practical demands of telecommunication companies, that seek to conduct cross-industry service innovation projects. It also generates new insights and propositions for existing theoretical innovation models. However, its clearly defined subject of analysis limits its generalization to other contexts what suggests further research.

First of all, the case study selection focusses on projects with up to three at least middle sized companies as innovation partner, and services in terms of intangible, reproducible equivalents of goods (Segelod & Jordan 2004) (cf. chapter 3.4.1). These requirements exclude customers, start-ups[5], universities and other research institutions, as well as customer specific developments, pure goods and pure services (Belderbos et al. 2006). Future research may produce valuable insights, if and how the deduced implications may alter, when they refuse these limitations (Gassmann, Enkel, et al. 2010). Moreover, the setting is dominated by a European telecommunication company. Even though the developed practical process model may serve as a basis for an innovation model across all continents, industries and companies, it needs to be verified to what extent it can be transferred. Consequently, future research should cover other industries or sectors, compare if the contributions vary depending on the focus industry or sector, and in addition, it may include cultural aspects and should analyze non-European environments impact on the contributions (Castellacci 2009; Alam 2006; Herranz 2006; Steenkamp et al. 1999).

Future studies may also improve and specify the present studies contributions, e.g., related to the emphasized importance of pilot customer and innovation partner, the suggested anchoring of the innovation project within the organizational structure, and the pursued integration of the new solution into the telco's product portfolio. Earlier studies reveal that the willingness of supplier to collaborate in innovation projects is determined by monetary compensation, know-how transfer and reputational effects (Smals & Smits 2012). This suggests to analyze if the same factors determine the willingness for potential pilot customers and innovation partner to participate in a cross-industry innovation project. The research may result in recommendations how to better identify and integrate respective pilot customers and innovation partners. Moreover, further insights regarding optimal size, legal organization and constellation of cross-industry innovation networks will be advantageous. Previous research identify that a short average path lengths to a wide range of firms within an alliance

---

[5] Hub:raum, the incubator of Deutsche Telekom and Optus-Innov8 Seed, the Optus and SingTel group's corporate venture capital team, are two exemplary options, that represent possible approaches to develop innovative ideas in more open, autarkic acting innovation cells, with relatively less bureaucratic processes, high degrees of freedom and the potential for a later clean integration in the telecommunication company portfolio or other sources of commercialization.

network will result in a greater innovative output, while others in contrast, reveal its limitations (Tsou 2012b; Schilling & Phelps 2007; Zahra & George 2002; Dyer & Singh 1998). Absorptive capacity research analyzes an organizations ability to identify the value of new information, coordinate, assimilate, and apply it to commercial ends (Grant 1996; Cohen & Levinthal 1990). There are routines for knowledge management available, facilitating the transfer of learning and practices across industries (Tranfield et al. 2003). Linked with insights from the absorptive capacity research stream and related to the specifics of cross-industry innovation, it may add value to the suggested structural and organizational collaboration characteristics and potential commercialization efforts (Tsai 2009) (cf. chapter 7.2 and 7.4).

One case study (cf. chapter 4.2) reveals, that even though an innovation project did not reach the desired results while developing the technical solution, it still was valued as a successful project, because it improved the relationship to the partner company and the telecommunication company gained extensive industry specific expertise. Innovation is very complex and multidimensional in nature, hence, the degree of innovativeness and success measures may differ according to the context (Chiesa et al. 2009). Soft factors may be one measure criterion, but more importantly they should be analyzed both on a technological as well as on a commercial level, however, there is still a lack of good and meaningful measures of innovation (Kleinknecht & Van Der Panne 2012; Goswami & Mathew 2005; Gemünden et al. 1992). Defining respective measures should be in focus of future research, because it will not just allow judging an innovation project ex post, but also continuously, what will make it possible to terminate unnecessary or unprofitable developments before they waste enormous amounts of innovation budget. As long as objective, timely and reliable measure are not available, and an organization feels pressure to pursue a specific solution for entering a specific market, it may be a good idea to develop two or more projects in parallel focusing on the same outcome (Kale & Singh 2009). A good management of a portfolio of alliances and of a portfolio of innovation projects allows an overarching risk management, increasing the chance that at least one solution reaches maturity, and consequently, accelerates the needed time-to-market to generate value (Kock et al. 2012; Koppinen et al. 2010; Hoffmann 2007). However, parallel development also increases the financial risk significantly and potentially needed experts, partner and customers might not be available as required. Future research should analyze the possibilities of conducting parallel multiple cross-industry innovation projects, to realize synergies, to guarantee quick time-to-market and to increase the probability of market success.

REFERENCES

Adner, R. & Kapoor, R., 2010. Value creation in innovation ecosystems: how the structure of technological interdependence affects firm performance in new technology generations. *Strategic Management Journal*, 31(3), pp.306–333.

Alam, I., 2007. New Service Development Process: Emerging versus Developed Markets. *Journal of Global Marketing*, 20(2/3), pp.43–55.

Alam, I., 2006. Service innovation strategy and process: a cross-national comparative analysis. *International Marketing Review*, 23(3), pp.234–254.

Alam, I. & Perry, C., 2002. A customer-oriented new service development process. *Journal of Services Marketing*, 16(6), pp.515–534.

Athaide, G.A. & Zhang, J.Q., 2011. The Determinants of Seller-Buyer Interactions during New Product Development in Technology-Based Industrial Markets. *Journal of Product Innovation Management*, 28(s1), pp.146–158.

Bach, V. & Whitehill, M., 2008. The profit factor: how corporate culture affects a joint venture. *Strategic Change*, 17(3/4), pp.115–132.

Baker, W.E. & Sinkula, J.M., 2007. Does market orientation facilitate balanced innovation programs? An organizational learning perspective. *Journal of Product Innovation Management*, 24(4), pp.316–334.

Baldwin, C. & von Hippel, E., 2011. Modeling a Paradigm Shift: From Producer Innovation to User and Open Collaborative Innovation. *Organization Science*, 22(6), pp.1399–1417.

Batterink, M.H. et al., 2010. Orchestrating innovation networks: The case of innovation brokers in the agri-food sector. *Entrepreneurship & Regional Development*, 22(1), pp.47–76.

Belderbos, R., Carree, M. & Lokshin, B., 2006. Complementarity in R&D Cooperation Strategies. *Review of Industrial Organization*, 28(4), pp.401–426.

Beringer, C., Jonas, D. & Gemünden, H.G., 2012. Establishing Project Portfolio Management: An Exploratory Analysis of the Influence of Internal Stakeholders' Interactions. *Project Management Journal*, 43(6), pp.16–32.

Bhattacharya, S. & Krishnan, V., 1998. Managing New Product Definition in Highly Dynamic Environments. *Management Science*, 44(11), pp.S50–S64.

Blomqvist, K. et al., 2004. Towards networked R&D management: the R&D approach of Sonera Corporation as an example. *R&D Management*, 34(5), pp.591–603.

Bluhm, D.J. et al., 2011. Qualitative Research in Management: A Decade of Progress. *Journal of Management Studies*, 48(8), pp.1866–1891.

Bond, E.U. & Houston, M.B., 2003. Barriers to Matching New Technologies and Market Opportunities in Established Firms. *Journal of Product Innovation Management*, 20(2), pp.120–135.

Brunswicker, S. & Hutschek, U., 2010. Crossing Horizons: Leveraging Cross-Industry Innovation Search in the Front-End of the Innovation Process. *International Journal of Innovation Management*, 14(04), pp.683–702.

Bstieler, L., 2006. Trust Formation in Collaborative New Product Development. *Journal of Product Innovation Management*, 23(1), pp.56–72.

Buganza, T. et al., 2011. Organisational implications of open innovation: An analysis of inter-industry patterns. *International Journal of Innovation Management*, 15(02), pp.423–455.

Burgelman, R.A., 1983. A Process Model of Internal Corporate Venturing in the Diversified Major Firm. *Administrative Science Quarterly*, 28(2), pp.223–244.

Calida, B. & Hester, P., 2010. Unraveling future research: an analysis of emergent literature in open innovation. *Annals of Innovation & Entrepreneurship*, 1(1). Available at: http://journals.sfu.ca/coaction/index.php/aie/article/view/5845 [Accessed October 25, 2012].

Cansfield, M., 2009. Farewell To The Traditional Telecom Ecosystem. Forrester Research, Inc.

Carbonell, P., Rodríguez-Escudero, A.I. & Pujari, D., 2009. Customer Involvement in New Service Development: An Examination of Antecedents and Outcomes. *Journal of Product Innovation Management*, 26(5), pp.536–550.

Cassell, C. et al., 2006. The role and status of qualitative methods in management research: an empirical account. *Management Decision*, 44(2), pp.290–303.

Castellacci, F., 2009. The interactions between national systems and sectoral patterns of innovation. *Journal of Evolutionary Economics*, 19(3), pp.321–347.

Chai, K.-H., Zhang, J. & Tan, K.-C., 2005. A TRIZ-Based Method for New Service Design. *Journal of Service Research : JSR*, 8(1), pp.48–66.

Chapas, R. et al., 2010. Sustainability in R&D. *Research Technology Management*, 53(6), pp.60–63.

Du Chatenier, E. et al., 2010. Identification of competencies for professionals in open innovation teams. *R&D Management*, 40(3), pp.271–280.

Chen, T.F., 2011. Building an integrated service innovation Model: A case study of Investment Banking. In *Economics, Trade and Development*. pp. 49–53.

Chesbrough, H., 2010. *Open services innovation: Rethinking your business to grow and compete in a new era*, John Wiley & Sons.

Chesbrough, H. & Euchner, J., 2011. Open Services Innovation: An Interview with Henry Chesbrough. *Research-Technology Management*, 54(2), pp.12–17.

Chesbrough, H.W., 2006. *Open business models : how to thrive in the new innovation landscape*, Boston: Harvar Business School.

Chesbrough, H.W. & Appleyard, M.M., 2007. Open innovation and strategy. *California management review*, 50(1), pp.57–76.

Chiaroni, D., Chiesa, V. & Frattini, F., 2011. The Open Innovation Journey: How firms dynamically implement the emerging innovation management paradigm. *Technovation*, 31(1), pp.34–43.

Chiaroni, D., Chiesa, V. & Frattini, F., 2010. Unravelling the process from Closed to Open Innovation: evidence from mature, asset-intensive industries. *R&D Management*, 40(3), pp.222–245.

Chien, S.-H. & Chen, J., 2010. Supplier involvement and customer involvement effect on new product development success in the financial service industry. *The Service Industries Journal*, 30(2), pp.185–201.

Chiesa, V. et al., 2009. Performance measurement in R&D: exploring the interplay between measurement objectives, dimensions of performance and contextual factors. *R&D Management*, 39(5), pp.487–519.

Chiesa, V. & Frattini, F., 2011. Commercializing Technological Innovation: Learning from Failures in High-Tech Markets*. *Journal of Product Innovation Management*, 28(4), pp.437–454.

Christensen, C. & Euchner, J., 2011. Managing disruption: an interview with Clayton Christensen. *Research-Technology Management*, 54(1), pp.11–17.

Christensen, C.M. & Roth, S.D.A.E.A., 2001. *Innovation in the Telecommunications Industry*, Telecommunications working paper.

Cohen, W.M. & Levinthal, D.A., 1990. Absorptive Capacity: A New Perspective on Learning and Innovation. *Administrative Science Quarterly*, 35(1), pp.128–152.

Conway, S., 1995. Informal boundary-spanning communication in the innovation process: an empirical study. *Technology Analysis & Strategic Management*, 7(3), pp.327–342.

Cooper, R.G., 1990. Stage-gate systems: A new tool for managing new products. *Business Horizons*, 33(3), pp.44–54.

Cooper, R.G. & Edgett, S.J., 2012. Best Practices in the Idea-to-Launch Process and Its Governance. *Research-Technology Management*, 55(2), pp.43–54.

Couchman, P.K. & Fulop, L., 2009. Examining partner experience in cross-sector collaborative projects focused on the commercialization of R&D. *Innovation: Management, Policy & Practice*, 11(1), pp.85–103.

Damanpour, F., 1996. Organizational complexity and innovation: Developing and testing multiple contingency models. *Management Science*, 42(5), p.693.

DeSanctis, G., Glass, J.T. & Ensing, I.M., 2002. Organizational designs for R&D. *The Academy of Management Executive*, 16(3), pp.55–66.

Dubé, L. & Paré, G., 2003. Rigor in information systems positivist case research: Current practices, trends, and recommendations. *MIS Quarterly*, 27(4), pp.597–635.

Dunne, C., 2011. The place of the literature review in grounded theory research. *International Journal of Social Research Methodology*, 14(2), pp.111–124.

Dyer, J.H. & Singh, H., 1998. The Relational View: Cooperative Strategy and Sources of Interorganizational Competitive Advantage. *Academy of Management Review*, 23(4), pp.660–679.

Eisenhardt, K.M., 1989. Building Theories From Case Study Research. *Academy of Management. The Academy of Management Review*, 14(4), p.532.

Eisenhardt, K.M. & Tabrizi, B.N., 1995. Accelerating Adaptive Processes: Product Innovation in the Global Computer Industry. *Administrative Science Quarterly*, 40(1), pp.84–110.

Eisingerich, A.B. & Bell, S.J., 2008. Managing networks of interorganizational linkages and sustainable firm performance in business-to-business service contexts. *Journal of Services Marketing*, 22(7), pp.494–504.

Engelen, A. & Brettel, M., 2012. A Coalitional Perspective on the Role of the R&D Department within the Organization. *Journal of Product Innovation Management*, 29(3), pp.489–505.

Enkel, E. & Gassmann, O., 2010. Creative imitation: exploring the case of cross-industry innovation. *R&D Management*, 40(3), pp.256–270.

Enkel, E., Gassmann, O. & Chesbrough, H., 2009. Open R&D and open innovation: exploring the phenomenon. *R&D Management*, 39(4), pp.311–316.

Enkel, E. & Goel, S., 2012. Smoothing the corporate venturing path: rules still count. *Journal of Business Strategy*, 33(3), pp.30–39.

Eunjung Lee, Mishna, F. & Brennenstuhl, S., 2010. How to Critically Evaluate Case Studies in Social Work. *Research on Social Work Practice*, 20(6), pp.682–689.

European Commission & Open Innovation Strategy and Policy Group, 2011. *Service innovation yearbook 2010-2011.*, Luxembourg: Publications Office of the European Union.

Evanschitzky, H. et al., 2012. Success Factors of Product Innovation: An Updated Meta-Analysis. *Journal of Product Innovation Management*, 29, pp.21–37.

Eveleens, C., 2010. Innovation management; a literature review of innovation process models and their implications. *Science*, 800, p.900.

Floricel, S. & Miller, R., 2003. An exploratory comparison of the management of innovation in the New and Old economies. *R&D Management*, 33(5), p.501.

Foxall, G. & Johnston, B., 1987. Strategies of user-initiated product innovation. *Technovation*, 6(2), pp.77–102.

Fransman, M., 2002. Mapping the evolving telecoms industry: the uses and shortcomings of the layer model. *Telecommunications Policy*, 26(9–10), pp.473–483.

Fraser, H., Mounib, E.L. & Payne, S., 2007. Cross-industry collaboration: a critical step to better serve patients. *hfm (Healthcare Financial Management)*, 61(12), pp.90–92.

Fredberg, T., Elmquist, M. & Ollila, S., 2008. *Managing open innovation: present findings and future directions*, Stockholm: VINNOVA.

Frisanco, T. et al., 2008. Telecommunications project management - a holistic approach for operations-related services. In *IEEE International Conference on Industrial Engineering and Engineering Management, 2008. IEEM 2008.* IEEE International Conference on Industrial Engineering and Engineering Management, 2008. IEEM 2008. pp. 1043–1048.

Froehle, C.M. & Roth, A.V., 2007. A Resource-Process Framework of New Service Development. *Production and Operations Management*, 16(2), pp.169–181,183–188.

Galanakis, K., 2006. Innovation process. Make sense using systems thinking. *Technovation*, 26(11), pp.1222–1232.

Gales, L. & Mansour-Cole, D., 1995. User involvement in innovation projects: Toward an information processing model. *Journal of Engineering and Technology Management*, 12(1–2), pp.77–109.

Garrette, B., Castañer, X. & Dussauge, P., 2009. Horizontal alliances as an alternative to autonomous production: product expansion mode choice in the worldwide aircraft industry 1945–2000. *Strategic Management Journal*, 30(8), pp.885–894.

Garrett-Jones, S. et al., 2005. Common purpose and divided loyalties: the risks and rewards of cross-sector collaboration for academic and government researchers. *R&D Management*, 35(5), pp.535–544.

Gassmann, O., Zeschky, M., et al., 2010. Crossing the Industry-Line: Breakthrough Innovation through Cross-Industry Alliances with "Non-Suppliers." *Long Range Planning*, 43(5–6), pp.639–654.

Gassmann, O., 2006. Opening up the innovation process: towards an agenda. *R&D Management*, 36(3), pp.223–228.

Gassmann, O., Daiber, M. & Enkel, E., 2011. The role of intermediaries in cross-industry innovation processes. *R&D Management*, 41(5), pp.457–469.

Gassmann, O. & Enkel, E., 2006. Open innovation. *ZfO Wissen*, 3(75), pp.132–138.

Gassmann, O. & Enkel, E., 2004. Towards a theory of open innovation: three core process archetypes. In *R&D management conference*. pp. 1–18. Available at: http://www.alexandria.unisg.ch/export/DL/20417.pdf [Accessed October 24, 2012].

Gassmann, O., Enkel, E. & Chesbrough, H., 2010. The future of open innovation. *R&D Management*, 40(3), pp.213–221.

Gassmann, O., Kausch, C. & Enkel, E., 2010. Negative side effects of customer integration. *International Journal of Technology Management*, 50(1), pp.43–63.

Gassmann, O. & Zeschky, M., 2008. Opening up the Solution Space: The Role of Analogical Thinking for Breakthrough Product Innovation. *Creativity and Innovation Management*, 17(2), pp.97–106.

Gemünden, H.G., Heydebreck, P. & Herden, R., 1992. Technological interweavement: a means of achieving innovation success. *R&D Management*, 22(4), pp.359–376.

Gemünden, H.G., Salomo, S. & Hölzle, K., 2007. Role Models for Radical Innovations in Times of Open Innovation. *Creativity and Innovation Management*, 16(4), pp.408–421.

Getha-Taylor, H., 2012. Cross-Sector Understanding and Trust. *Public Performance & Management Review*, 36(2), pp.216–229.

Gibbs, G.R., Friese, S. & Mangabeira, W.C., 2002. The Use of New Technology in Qualitative Research. Introduction to Issue 3(2) of FQS. *Forum: Qualitative Social Research*, 3(2). Available at: http://search.proquest.com.ezproxy2.library.usyd.edu.au/docview/867760 319/abstract/139F6ACA8A26E7B4976/1?accountid=14757 [Accessed October 24, 2012].

Gillier, T. et al., 2010. Managing Innovation Fields in a Cross-Industry Exploratory Partnership with C–K Design Theory. *Journal of product innovation management*, 27(6), pp.883–896.

Gillier, T., Kazakci, A.O. & Piat, G., 2012. The generation of common purpose in innovation partnerships. *European Journal of Innovation Management*, 15(3), pp.372–392.

Giuri, P. et al., 2007. Inventors and invention processes in Europe: Results from the PatVal-EU survey. *Research Policy*, 36(8), pp.1107–1127.

Glaser, B.G. & Strauss, A.L., 1967. *The discovery of grounded theory: strategies for qualitative research*, New York: Aldine De Gruyter.

Golafshani, N., 2003. Understanding reliability and validity in qualitative research. *The qualitative report*, 8(4), pp.597–607.

Goswami, S. & Mathew, M., 2005. Definition of Innovation Revisited: An Empirical Study on Indian Information Technology Industry. *International Journal of Innovation Management*, 9(3), pp.371–383.

Gottfridsson, P., 2012. Joint service development – the creations of the prerequisite for the service development. *Managing Service Quality*, 22(1), pp.21–37.

Granovetter, M., 2005. The impact of social structure on economic outcomes. *The Journal of Economic Perspectives*, 19(1), pp.33–50.

Granovetter, M.S., 1973. The Strength of Weak Ties. *American Journal of Sociology*, 78(6), pp.1360–1380.

Grant, R.M., 1996. Toward a Knowledge-Based Theory of the Firm. *Strategic Management Journal*, 17, pp.109–122.

Grant, R.M. & Baden-Fuller, C., 2003. A Knowledge Accessing Theory of Strategic Alliances. *Journal of Management Studies*, 41(1), pp.61–84.

Grönlund, J., Sjödin, D.R. & Frishammar, J., 2010. Open innovation and the stage-gate process: A revised model for new product development. *California Management Review*, 52(3), pp.106–131.

Grote, M., Herstatt, C. & Gemünden, H.G., 2012. Cross-Divisional Innovation in the Large Corporation: Thoughts and Evidence on Its Value and the Role of the Early Stages of Innovation. *Creativity & Innovation Management*, 21(4), pp.361–375.

Gupta, A., Pawar, K.S. & Smart, P., 2007. New product development in the pharmaceutical and telecommunication industries: A comparative study. *International Journal of Production Economics*, 106(1), pp.41–60.

Hagedoorn, J. & Duysters, G., 2002. External Sources of Innovative Capabilities: The Preferences for Strategic Alliances or Mergers and Acquisitions. *Journal of Management Studies*, 39(2), pp.167–188.

Han, K. et al., 2012. Value cocreation and wealth spillover in open innovation alliances. *MIS Quarterly-Management Information Systems*, 36(1), p.291.

Hansen, J.D. & Bunn, M.D., 2009. Stakeholder Relationship Management in Multi-Sector Innovations. *Journal of Relationship Marketing*, 8(3), pp.196–217.

Harmancioglu, N. et al., 2007. Your new product development (NPD) is only as good as your process: an exploratory analysis of new NPD process design and implementation. *R&D Management*, 37(5), pp.399–424.

Häussler, C., 2010. The economics of knowledge regulation: an empirical analysis of knowledge flows. *R&D Management*, 40(3), pp.300–309.

Heath, H. & Cowley, S., 2004. Developing a grounded theory approach: a comparison of Glaser and Strauss. *International Journal of Nursing Studies*, 41(2), pp.141–150.

Herranz, J., 2006. The Multisectoral Trilemma of Network Management. *Journal of Public Administration Research and Theory*, 18(1), pp.1–31.

Hess, T. et al., 2012. IKT-Anbieter als Thema der Wirtschaftsinformatik? *WIRTSCHAFTSINFORMATIK*, 54(6), pp.355–362.

Hill, C.W.L. & Rothaermel, F.T., 2003. The Performance of Incumbent Firms in the Face of Radical Technological Innovation. *Academy of Management Review*, 28(2), pp.257–274.

Von Hippel, E., 1976. The dominant role of users in the scientific instrument innovation process. *Research Policy*, 5(3), pp.212–239.

Hirst, G. & Mann, L., 2004. A model of R&D leadership and team communication: the relationship with project performance. *R&D Management*, 34(2), pp.147–160.

Hoegl, M. & Gemuenden, H.G., 2001. Teamwork Quality and the Success of Innovative Projects: A Theoretical Concept and Empirical Evidence. *Organization Science*, 12(4), pp.435–449.

Hoegl, M., Weinkauf, K. & Gemuenden, H.G., 2004. Interteam Coordination, Project Commitment, and Teamwork in Multiteam R&D Projects: A Longitudinal Study. *Organization Science*, 15(1), pp.38–55.

Hoffmann, W.H., 2007. Strategies for managing a portfolio of alliances. *Strategic Management Journal*, 28(8), pp.827–856.

Howells, J., 2006. Intermediation and the role of intermediaries in innovation. *Research Policy*, 35(5), pp.715–728.

Howells, J. & Tether, B., 2004. Innovation in services: Issues at stake and trends. *INNO-Studies 2001: Lot 3 (ENTR-C/2001), Brussels.*

Hsieh, K.-N. & Tidd, J., 2012. Open versus closed new service development: The influences of project novelty. *Technovation*, 32(11), pp.600–608.

Hummel, E. et al., 2010. Business Models for Collaborative Research. *Research Technology Management*, 53(6), pp.51–54.

Huston, L. & Sakkab, N., 2006. Connect and develop. *Harvard business review*, 84(3), pp.58–66.

Iansiti, M. & West, J., 1997. Technology Integration: Turning Great Research into Great Products. *Harvard Business Review*, 75(3), pp.69–79.

Idrees, I., Vasconcelos, A.C. & Cox, A.M., 2011. The use of Grounded Theory in PhD research in knowledge management. *Aslib Proceedings*, 63(2/3), pp.188–203.

Inauen, M. & Schenker-Wicki, A., 2012. Fostering radical innovations with open innovation. *European Journal of Innovation Management*, 15(2), pp.212–231.

Inauen, M. & Schenker-Wicki, A., 2011. The impact of outside-in open innovation on innovation performance. *European Journal of Innovation Management*, 14(4), pp.496–520.

Jiménez-Zarco, A.I., Martínez-Ruiz, M.P. & Izquierdo-Yusta, A., 2011. The impact of market orientation dimensions on client cooperation in the development of new service innovations. *European Journal of Marketing*, 45(1/2), pp.43–67.

Johannessen, J.-A., Olsen, B. & Lumpkin, G.T., 2001. Innovation as newness: what is new, how new, and new to whom? *European Journal of innovation management*, 4(1), pp.20–31.

Kale, P. & Singh, H., 2009. Managing strategic alliances: what do we know now, and where do we go from here? *The Academy of Management Perspectives*, 23(3), pp.45–62.

Kale, P., Singh, H. & Perlmutter, H., 2000. Learning and Protection of Proprietary Assets in Strategic Alliances: Building Relational Capital. *Strategic Management Journal*, 21(3), pp.217–237.

Kindström, D. & Kowalkowski, C., 2009. Development of industrial service offerings: a process framework. *Journal of Service Management*, 20(2), pp.156–172.

Kleinknecht, A. & Van Der Panne, G., 2012. Predicting New Product Sales:: The Post-Launch Performance of 215 Innovators. *International Journal of Innovation Management*, 16(2), pp.1250011–1.

Kock, A., Gemünden, H.G. & Jonas, D., 2012. Wertsteigerung durch Projekt portfoliomanagement. (German). *Zeitschrift Fü;hrung und Organisation*, 81(1), pp.4–9.

Koppinen, S., Lammasniemi, J. & Kalliokoski, P., 2010. Practical application of a parallel research–business innovation process to accelerate the deployment of research results. *R&D Management*, 40(1), pp.101–106.

Kotabe, M. & Swan, K.S., 1995. The Role of Strategic Alliances in High-Technology New Product Development. *Strategic Management Journal*, 16(8), pp.621–636.

Lau, A.K.W., Tang, E. & Yam, R.C.M., 2010. Effects of Supplier and Customer Integration on Product Innovation and Performance: Empirical Evidence in Hong Kong Manufacturers. *Journal of Product Innovation Management*, 27(5), pp.761–777.

Lazzarotti, V. & Manzini, R., 2009. Different modes of open innovation: a theoretical framework and an empirical study. *International journal of innovation management*, 13(04), pp.615–636.

Lee, T.W., Mitchell, T.R. & Sablynski, C.J., 1999. Qualitative Research in Organizational and Vocational Psychology, 1979–1999. *Journal of Vocational Behavior*, 55(2), pp.161–187.

Leimbach, T. & Friedewald, M., 2010. Assessing national policies to support software in Europe. *Info : the Journal of Policy, Regulation and Strategy for Telecommunications, Information and Media*, 12(6), pp.40–55.

Leiponen, A. & Drejer, I., 2007. What exactly are technological regimes?: Intra-industry heterogeneity in the organization of innovation activities. *Research Policy*, 36(8), pp.1221–1238.

Liao, Z., 2001. International R&D project evaluation by multinational corporations in the electronics and IT industry of Singapore. *R&D Management*, 31(3).

Lichtenthaler, U., 2011. Open Innovation: Past Research, Current Debates, and Future Directions. *The Academy of Management Perspectives*, 25(1), pp.75–93.

Li, F. & Whalley, J., 2002. Deconstruction of the telecommunications industry: from value chains to value networks. *Telecommunications Policy*, 26(9–10), pp.451–472.

Linnarson, H., 2005. Patterns of alignment in alliance structure and innovation. *Technology Analysis & Strategic Management*, 17(2), pp.161–181.

Lovelock, C. & Gummesson, E., 2004. Whither Services Marketing? In Search of a New Paradigm and Fresh Perspectives. *Journal of Service Research*, 7(1), pp.20–41.

Lynn, G., Morone, J. & Paulson, A., 1996. Marketing and discontinuous innovation: the probe and learn process. *California management review*, 38(3). Available at: http://papers.ssrn.com/sol3/papers.cfm?abstract_id=2151914 [Accessed June 1, 2013].

Magnusson, P.R., 2003. Benefits of involving users in service innovation. *European Journal of Innovation Management*, 6(4), pp.228–238.

Maklan, S., Knox, S. & Ryals, L., 2008. New trends in innovation and customer relationship management. *International Journal of Market Research*, 50(2), pp.221–240.

Mann, D., 2001. An Introduction to TRIZ: The Theory of Inventive Problem Solving. *Creativity & Innovation Management*, 10(2), p.123.

Mansfeld, M.N., Hölzle, K. & Gemünden, H.G., 2010. Personal Characteristics of Innovators - An empirical study of roles in innovation management. *International Journal of Innovation Management*, 14(6), pp.1129–1147.

Marinova, D., 2004. Actualizing Innovation Effort: The Impact of Market Knowledge Diffusion in a Dynamic System of Competition. *Journal of Marketing*, 68(3), pp.1–20.

Martin, C.R.J. & Horne, D.A., 1995. Level of success inputs for service innovations in the same firm. *International Journal of Service Industry Management*, 6(4), p.40.

McGuinness, N.W. & Conway, H.A., 1989. Managing the search for new product concepts: a strategic approach. *R&D Management*, 19(4), pp.297–308.

Miles, M.B. & Huberman, A.M., 1994. *Qualitative data analysis: an expanded sourcebook* 2nd ed., Thousand Oaks: Sage Publications.

Miles, R.E., Miles, G. & Snow, C.C., 2006. Collaborative Entrepreneurship: A Business Model for Continuous Innovation. *Organizational Dynamics*, 35(1), pp.1–11.

Miller, R. & Olleros, X., 2007. The Dynamics of Games of Innovation. *International Journal of Innovation Management*, 11(1), pp.37–64.

Mishra, A.A. & Shah, R., 2009. In union lies strength: Collaborative competence in new product development and its performance effects. *Journal of Operations Management*, 27(4), pp.324–338.

Moehrle, M.G., 2005. What is TRIZ? From Conceptual Basics to a Framework for Research. *Creativity & Innovation Management*, 14(1), pp.3–13.

Mortara, L. et al., 2009. *How to implement open innovation: lessons from studying large multinational companies*, Cambridge: University of Cambridge Institute for Manufacturing.

Nambisan, S., Bacon, J. & Throckmorton, J., 2012. The Role of the Innovation Capitalist in Open Innovation. *Research Technology Management*, 55(3), pp.49–57.

Nayar, V., 2011. Meeting the Decade's Challenges: Technology (Alone) Is Not the Answer. In S. Dutta & I. Mia, eds. *The global information technology report 2010-2011: Transformations 2.0*. World Economic Forum and INSEAD.

Nesse, P., 2008. Open service innovation in telecom industry-case study of partnership models enabling 3rd party development of novel mobile services. *ICIN conference, Bordeaux, France*. Available at: https://www.icin.biz/files/2008papers/Session8A-2.pdf [Accessed October 25, 2012].

Newman, J.L., 2009. Building a Creative High-Performance R&d Culture. *Research Technology Management*, 52(5), pp.21–31.

Niosi, J., 1999. Fourth-generation R&D: From linear models to flexible innovation. *Journal of business research*, 45(2), pp.111–117.

Nobelius, D., 2004. Towards the sixth generation of R&D management. *International Journal of Project Management*, 22(5), pp.369–375.

Nolf, B., Tsiakis, P. & Sambukumar, R., 2012. How to Implement an Effective Market Scan. *Supply & Demand Chain Executive*, 13(1), pp.29–30.

Nooteboom, B. et al., 2007. Optimal cognitive distance and absorptive capacity. *Research Policy*, 36(7), pp.1016–1034.

OECD & Eurostat, 2005. *Oslo Manual: Guidelines for collecting and interpreting innovation data*. 3rd Edition., Paris: Organisation for Economic Co-operation and Development: Statistical Office of the European Communities.

OECD & Eurostat, 1997. *Oslo Manual: Proposed Guidelines for Collecting and Interpreting Technological Innovation Data* 2nd Edition., Organisation for Economic Co-operation and Development: Statistical Office of the European Communities.

Ordanini, A. & Parasuraman, A., 2011. Service Innovation Viewed Through a Service-Dominant Logic Lens: A Conceptual Framework and Empirical Analysis. *Journal of Service Research*, 14(1), pp.3–23.

Österle, H., Brenner, W. & Hilbers, K., 1991. *Unternehmensführung und Informationssystem: Der Ansatz des St. Galler Informationssystem-Managements*, Walter Brenner.

Ottenbacher, M.C. & Harrington, R.J., 2010. Strategies for achieving success for innovative versus incremental new services. *Journal of Services Marketing*, 24(1), pp.3–15.

Perks, H. & Riihela, N., 2004. An exploration of inter-functional integration in the new service development process. *Service Industries Journal*, 24(6), pp.37–63.

Plewa, C. et al., 2012. Technology adoption and performance impact in innovation domains. *Industrial Management & Data Systems*, 112(5), pp.748–765.

Du Preez, N.D. & Louw, L., 2008. A framework for managing the innovation process. In *Portland International Conference on Management of Engineering Technology, 2008. PICMET 2008*. Portland International Conference on Management of Engineering Technology, 2008. PICMET 2008. pp. 546–558.

Quinn, J.B. & Mueller, J.A., 1963. Transferring Research Results to Operations. *Harvard Business Review*, 41(1), pp.49–66.

Raivio, Y., Luukkainen, S. & Juntunen, A., 2009. Open Telco: a new business potential. In *Proceedings of the 6th International Conference on Mobile Technology, Application & Systems*. p. 2. Available at: http://dl.acm.org/citation.cfm?id=1710037 [Accessed October 25, 2012].

Rathmell, J.M., 1966. What Is Meant by Services? *Journal of Marketing*, 30(4), pp.32–36.

Ravenswood, K., 2011. Eisenhardt's impact on theory in case study research. *Journal of Business Research*, 64(7), pp.680–686.

Rese, A., Gemünden, H.-G. & Baier, D., 2013. "Too Many Cooks Spoil The Broth": Key Persons and their Roles in Inter-Organizational Innovations. *Creativity & Innovation Management*, 22(4), pp.390–407.

Ribiere, V.M. & Tuggle, F.D. (Doug), 2010. Fostering innovation with KM 2.0. *VINE*, 40(1), pp.90–101.

Riedl, R., 2006. Erkenntnisfortschritt durch Forschungsfallstudien - Überlegungen am Beispiel der Wirtschaftsinformatik. In S. Zelewski & N. Akca, eds. *Fortschritt in den Wirtschaftswissenschaften:*

wissenschaftstheoretische Grundlagen und exemplarische Anwendungen. Wiesbaden: Deutscher Universitäts-Verlag.

Riel, A.C.R. van et al., 2011. Technology-based service proposal screening and decision-making effectiveness. *Management Decision*, 49(5), pp.762–783.

Riel, A.C.R. van & Lievens, A., 2004. New service development in high tech sectors: A decision-making perspective. *International Journal of Service Industry Management*, 15(1), pp.72–101.

Riempp, G., 2004. *Integrierte Wissensmanagement-Systeme: Architektur und praktische Anwendung: mit 101 Abbildungen und 26 Tabellen*, Springer DE.

Ritala, P., Armila, L. & Blomqvist, K., 2009. Innovation orchestration capability—Defining the organizational and individual level determinants. *International Journal of Innovation Management*, 13(04), pp.569–591.

Roberts, E.B., 2007. Managing Invention and Innovation. *Research Technology Management*, 50(1), pp.35–54.

Rohrbeck, R., Hölzle, K. & Gemünden, H.G., 2009. Opening up for competitive advantage – How Deutsche Telekom creates an open innovation ecosystem. *R&D Management*, 39(4), pp.420–430.

Rothwell, R., 1992. Successful industrial innovation: critical factors for the 1990s. *R&D Management*, 22(3), pp.221–240.

Russo-Spena, T. & Mele, C., 2012. "Five Co-s" in innovating: a practice-based view. *Journal of Service Management*, 23(4), pp.527–553.

Salomo, S., Weise, J. & Gemünden, H.G., 2007. NPD Planning Activities and Innovation Performance: The Mediating Role of Process Management and the Moderating Effect of Product Innovativeness. *Journal of Product Innovation Management*, 24(4), pp.285–302.

Schilling, M.A. & Phelps, C.C., 2007. Interfirm Collaboration Networks: The Impact of Large-Scale Network Structure on Firm Innovation. *Management Science*, 53(7), pp.1113–1126.

Schreiner, M., Kale, P. & Corsten, D., 2009. What really is alliance management capability and how does it impact alliance outcomes and success? *Strategic Management Journal*, 30(13), pp.1395–1419.

Schweitzer, F. & Gabriel, I., 2012. Action at the front end of innovation. *International Journal of Innovation Management*, 16(06), p.1240010.

Segelod, E. & Jordan, G., 2004. The use and importance of external sources of knowledge in the software development process. *R&D Management*, 34(3), pp.239–252.

Shenton, A.K., 2004. Strategies for ensuring trustworthiness in qualitative research projects. *Education for information*, 22(2), pp.63–75.

Sienel, J. et al., 2009. OPUCE: A telco-driven service mash-up approach. *Bell Labs Technical Journal*, 14(1), pp.203–218.

Smals, R.G.M. & Smits, A.A.J., 2012. Value for value—The dynamics of supplier value in collaborative new product development. *Industrial Marketing Management*, 41(1), pp.156–165.

Song, L.Z., Song, M. & Di Benedetto, C.A., 2009. A Staged Service Innovation Model. *Decision Sciences*, 40(3), pp.571–599.

Song, M. et al., 2011. Does Strategic Planning Enhance or Impede Innovation and Firm Performance? *Journal of Product Innovation Management*, 28(4), pp.503–520.

Steenkamp, J.-B.E.M., Hofstede, F. ter & Wedel, M., 1999. A Cross-National Investigation into the Individual and National Cultural Antecedents of Consumer Innovativeness. *Journal of Marketing*, 63(2), pp.55–69.

Stevens, E. & Dimitriadis, S., 2004. New service development through the lens of organisational learning: evidence from longitudinal case studies. *Journal of Business Research*, 57(10), pp.1074–1084.

Stigler, H., 2005. *Praxisbuch empirische Sozialforschung in den Erziehungs- und Bildungswissenschaften*, Innsbruck; Wien; Bozen: Studien-Verl.

Strauss, A.L. & Corbin, J.M., 1998. *Basics of qualitative research: techniques and procedures for developing grounded theory*, Thousand Oaks: Sage Publications.

Su, C.-T., Lin, C.-S. & Chiang, T.-L., 2008. Systematic improvement in service quality through TRIZ methodology: An exploratory study. *Total Quality Management & Business Excellence*, 19(3), pp.223–243.

Su, Y.-S., Tsang, E. & Peng, M., 2009. How do internal capabilities and external partnerships affect innovativeness? *Asia Pacific Journal of Management*, 26(2), pp.309–331.

Teece, D., 1989. Inter-Organizational Requirements of the Innovation Process. *Managerial and Decision Economics*, pp.35–42.

Telstra Corporation Ltd. & KPMG International, 2012. Cross Industry Innovation 2012 – The secret may well be in another industry.

Terreberry, S., 1968. The evolution of organizational environments. *Administrative Science Quarterly*, pp.590–613.

Tether, B.S., 2002. Who co-operates for innovation, and why - An empirical analysis. *Research Policy*, 31(6), pp.947–967.

Thomke, S., 2001. Enlightened Experimentation: The New Imperative for Innovation. *Harvard Business Review*, 79(2), pp.67–75.

Thomke, S., 2003. R&D Comes to Services. *Harvard Business Review*, 81(4), pp.70–79.

Tranfield, D. et al., 2003. Knowledge Management Routines for Innovation Projects: Developing a Hierarchical Process Model. *International Journal of Innovation Management*, 7(1), p.27.

Trott, P. & Hartmann, D.A.P., 2009. Why'open innovation'is old wine in new bottles. *International Journal of Innovation Management*, 13(04), pp.715–736.

Tsai, K.-H., 2009. Collaborative networks and product innovation performance: Toward a contingency perspective. *Research Policy*, 38(5), pp.765–778.

Tsou, H.-T., 2012a. Collaboration competency and partner match for e-service product innovation through knowledge integration mechanisms. *Journal of Service Management*, 23(5), pp.640–663.

Tsou, H.-T., 2012b. The effect of interfirm codevelopment competency on the innovation of the e-service process and product: the perspective of internal/external technology integration mechanisms. *Technology Analysis & Strategic Management*, 24(7), pp.631–646.

Tsou, H.-T. & Chen, J.-S., 2012. The influence of interfirm codevelopment competency on e-service innovation. *Information & Management*, 49(3–4), pp.177–189.

Unger, B.N. et al., 2012. Enforcing strategic fit of project portfolios by project termination: An empirical study on senior management involvement. *International Journal of Project Management*, 30(6), pp.675–685.

Vanhaverbeke, W., Van de Vrande, V. & Chesbrough, H., 2008. Understanding the Advantages of Open Innovation Practices in Corporate Venturing in Terms of Real Options. *Creativity and Innovation Management*, 17(4), pp.251–258.

Vargo, S.L. & Lusch, R.F., 2004a. Evolving to a New Dominant Logic for Marketing. *Journal of Marketing*, 68(1), pp.1–17.

Vargo, S.L. & Lusch, R.F., 2004b. The Four Service Marketing Myths Remnants of a Goods-Based, Manufacturing Model. *Journal of Service Research*, 6(4), pp.324–335.

Voss, M., 2012. Impact of customer integration on project portfolio management and its success—Developing a conceptual framework. *International Journal of Project Management*, 30(5), pp.567–581.

Weck, M., 2006. Knowledge creation and exploitation in collaborative R&D projects: lessons learned on success factors. *Knowledge & Process Management*, 13(4), pp.252–263.

Wießmeier, G.F.L., Thoma, A. & Senn, C., 2012. Leveraging Synergies Between R&D and Key Account Management to Drive Value Creation. *Research Technology Management*, 55(3), pp.15–22.

Wilde, T. & Hess, T., 2007. Forschungsmethoden der Wirtschaftsinformatik. *Wirtschaftsinformatik*, 49(4), pp.280–287.

Wulf, J. & Zarnekow, R., 2011. Cross-Sector Competition in Telecommunications. *Business & Information Systems Engineering*, 3(5), pp.289–298.

Yamashina, H., Ito, T. & Kawada, H., 2002. Innovative product development process by integrating QFD and TRIZ. *International Journal of Production Research*, 40(5), pp.1031–1050.

Yang, C.-C., 2007. A Systems Approach to Service Development in a Concurrent Engineering Environment. *Service Industries Journal*, 27(5), pp.635–652.

Yin, R.K., 2003a. *Applications of case study research* Second Edition., Thousand Oaks: Sage Publications.

Yin, R.K., 2003b. *Case study research : design and methods* Third Edition., Thousand Oaks, Calif.: Sage Publications.

Yin, R.K., 2011. *Qualitative research from start to finish*, New York: Guilford Press.

Yin, R.K., 1981. The case study crisis: some answers. *Administrative science quarterly*, 26(1), pp.58–65.

Zahra, S.A. & George, G., 2002. Absorptive Capacity: A Review, Reconceptualization, and Extension. *Academy of Management Review*, 27(2), pp.185–203.

Zeithaml, V.A., Parasuraman, A. & Berry, L.L., 1985. Problems and Strategies in Services Marketing. *Journal of Marketing*, 49(2), pp.33–46.

Zeng, L., Proctor, R.W. & Salvendy, G., 2011. Fostering Creativity in Product and Service Development Validation in the Domain of Information Technology. *Human Factors: The Journal of the Human Factors and Ergonomics Society*, 53(3), pp.245–270.

APPENDIX

I. Interview Guideline

The following guideline was used in German language in all interviews. All questions behind the black bullets were mandatory, the shifted questions in parentheses have only been asked if needed:

Introduction (5 min.):

Reception, Background information about dissertation, Person, Use of interview material within dissertation, Duration of the interview, Point to voice recorder.

- Do you have any questions in advance?

[Start voice recorder]

- Please name your function, tasks/responsibilities within the project.

I.) Conditions – General information about the cross-industry service innovation project (15 min.):

- What was the name of the project and what is the name of the emerged service?
- When was the project carried out?
  - ○ (Start-/end time? Duration?)
- What was the scope/size of the project?
  - ○ (Budget? Man-days? Size project team?)
- What is the completed service about?
- Why was the project initiated and conducted at all?
  - ○ (Image reasons? Financial reasons? Further development of something existing?)
- Why did the external partners participate?
  - ○ (Financial? Knowledge exchange? Strategic partnership?)
- How was the project organized and what processes have been followed?
  - ○ (Project plan? Mutually created? Guidelines? Standard processes? Project office?
  - ○ (Note to focus on the innovation process in the following)

- Who (internal and external player) have been involved and how have they been identified and selected?
  - (Who took the leading role? Which competences have been brought in by the parties?)
- Have key success criteria been defined – for service, project, collaboration, etc.?
  - (Which? How tracked? How reviewed? Have milestones been defined?)

II.) The innovation process, its phases (e.g.: ideation, market/business analysis, design, test, launch) and collaboration among the partners (30 min.):

- Please describe which phases have been passed through within the project.
  - (In which order? Duration? Have there been overlapping? What was done precisely within the phases?)
- Please describe the collaboration within the phases. Please focus on the collaboration with the external partners from outside the telco industry.
  - (Who with whom? In which phases? How intensive?)
  - (Planned and official, or informal via personal contacts, consulted experts, etc.?)
  - (What kind of communication media? Personal informal, common meetings, e-mail, SharePoint, etc.)
  - (In respect to the collaboration: What have been the core competences of the telecommunication company? What has contributed the most to the realization of the project? Organization? Coordination? Provisioning of technical infrastructure?)
- (Why) was in some phases no or less collaboration with the external partners?
  - (Competencies? Privacy reasons? Different ways of working?)
- Was the innovation process and the kind of collaboration in this project different from projects usually realized within the company?
  - (How exactly within the phases? What was the reason?)

- o (Is it possible to distinguish in (1) traditional/closed, (2) open and (3) cross-industry service innovation projects?)
- o (Please describe the companies codified service innovation process (sequence of phases, collaboration with external/non-telco, etc.))
- o (What was new for the external partners? Process? Collaboration?)
- Do you know anything about innovation process and collaboration compared to national and international competitors?
  - o (How do planning and realization differ?)
  - o (Are there difficulties due to history or culture?)

III.) Room for improvement, implications und lessons learned (30 min.):

- From an Innovation Management perspective, what led to successful planning and implementation of the project? What could have been better?
  - o (Was the project successful at all? Measured by what? Criteria achieved?
  - o (What were the biggest obstacles within the single phases? Where did planning and implementation diverge the most? And why? Which incidents resulted in delays?)
  - o (What was hindered/prevented by internal bureaucratic obstacles? Which company regulation could hardly be followed?)
  - o (From your perspective: how do the external partners evaluate planning and implementation of the project? How do you constitute your assumption?)
- What led to good cross-industry collaboration with the external partners, respectively what could have been better?
  - o (Common milestones? Mutual reservation? Knowledge?)
  - o (Have there been situations where you did not want to intensify the collaboration? E.g., due to privacy reasons, mutual understanding, ways of working, etc.? What has been done to solve it? Have the external partners been in a similar situation?)
  - o (Would it have been possible to improve the communication? How? Prior to or during the project? Technical? Personal? Judicial?)
  - o (How did you personally like the exchange?)

- o (From your perspective: how do the external partners evaluate the collaboration within the project? How do you constitute your assumption?)
- What competences did the external partners bring in, that were essential for the success of the project? What did you expect in addition?
  - o (What should the external partners have brought in to a greater extent? Was it requested? Why did they not offer it on their own?)
  - o (Did the external partners offer something to bring in, but you did not want it? Why did they offer it?)
- What competences did the telecommunication company bring in, that were essential for the success of the project? What did the external partners expect in addition?
  - o (Did the external partners request anything that you did not offer right from the start? Organizational? Technical? Why did you not offer it on your own?)
  - o (Did you offer something to bring in to a greater extent, but the external partners did not want it? Why did they not want it?)
  - o (From your perspective: is there anything you did not bring in enough? How would the external partners react?)
- To sum up, what criteria need to be fulfilled, to improve the success of cross-industry service innovation projects?
  - o (How is it possible to implement it within each single phase? How to measure the success?)
  - o (How can be assured that it is going to be a successful project for both sides?)
  - o (Refer to what was earlier said:
    - ▪ I.): Standard processes, Team composition, Interests, Success criteria/milestones, etc.
    - ▪ II.) Process phases, Planning/Codification, Allocation of tasks, Communication media, Competition, etc.)
- Is there anything we did not mention so far, that could be relevant for the formation of service innovation projects with cross-industry partners?

End – Thank you very much for your time.